LIFE *and* MIND

LIFE *and* MIND

THE LIGHT OF SYSTEM PHILOSOPHY

SCIENCE AND PHILOSOPHY

GEORGE LUKE

Copyright © 2018 by George Luke.

ISBN:	Softcover	978-1-5437-0421-1
	eBook	978-1-5437-0420-4

All rights reserved. No part of this book may be used or reproduced by any means, graphic, electronic, or mechanical, including photocopying, recording, taping or by any information storage retrieval system without the written permission of the author except in the case of brief quotations embodied in critical articles and reviews.

Because of the dynamic nature of the Internet, any web addresses or links contained in this book may have changed since publication and may no longer be valid. The views expressed in this work are solely those of the author and do not necessarily reflect the views of the publisher, and the publisher hereby disclaims any responsibility for them.

First published by PGL BOOKS as eBook in Amazon Kindle Store

Print information available on the last page.

To order additional copies of this book, contact
Partridge India
000 800 10062 62
orders.india@partridgepublishing.com

www.partridgepublishing.com/india

Also by GEORGE LUKE

SAPTALOKADARSHANAM SAMGRAHAM (MALAYALAM)
JEEVANUM PARINAMAVUM – SYSTEM PHILOSOPHIYUDE VELICHAM
      (MALAYALAM)
ORIGIN OF UNIVERSE : THE LIGHT OF SYSTEM PHILOSOPHY
DISCOVERY OF REALITY : THE LIGHT OF SYSTEM PHILOSOPHY

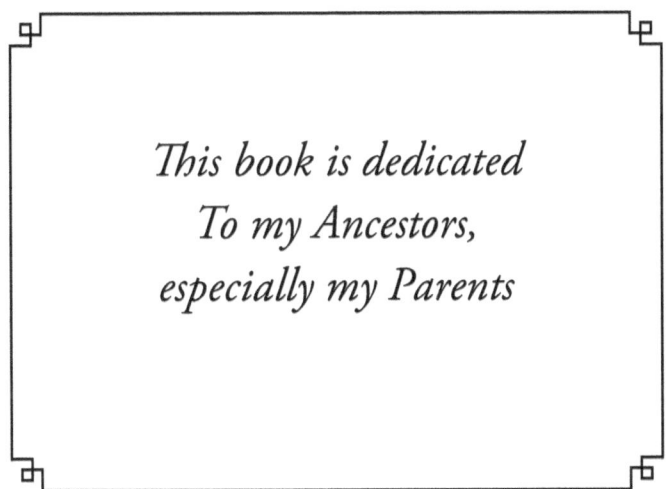

*This book is dedicated
To my Ancestors,
especially my Parents*

CONTENTS

Preface ix

Prologue xv

Introduction xxiii

Chapter 1: Key Ideas from Previous Book 1

    1.1 The Framework of Physical Science 1

    1.2 Dilemmas about the Origin of Universe 4

    1.3 Philosophy of Science and its Failures 5

    1.4 Vision of System Philosophy of Science 8

Chapter 2: Existence of Life 13

    2.1 The Puzzle of Life 14

    2.2 Modern Biology 22

    2.3 Criticism on Computer Model of Life 29

    2.4 System Philosophy of Life 34

        2.4.1 Philosophical Definition of System 36

        2.4.2 Cell is a Phenomenal System 38

Chapter 3: System Philosophy of Evolution 45

    3.1 Organic and Mechanistic Worldviews 47

    3.2 Spiritual Process View of Evolution 47

    3.3 Physical Process View : Darwinism 49

    3.4 Fallacy of Intelligent Designer Argument 55

    3.5 The System Model of Biological Evolution.......................58

Chapter 4: System Philosophy of Mind 69

    4.1 Human Mind and Nervous System 71

    4.2 Conflicting Philosophical Views about Human Mind 77

    4.3 System Philosophy of Mind...85

        4.3.1 Structure of Human Mind: The System Model...........88

        4.3.2 Solution of Mind-Body Dualism 96

        4.3.3 What is Consciousness? 102

Chapter 5: System Philosophy of God and Evil 108

    5.1 Anatomy of a Crisis 109

    5.2 Religion is a Social System.. 112

    5.3 God and Soul are Mystical Concepts................................. 114

    5.4 Justification of Religious Knowledge: Does God Exist?...... 117

Chapter 6: Science-Religion Synthesis 124

    6.1 Atheism is a Delusion 125

    6.2 Synthesis of Science and Religion 127

Bibliography...131

Index of Names.. 139

Index of Subjects 143

Preface

There are obvious categorical differences between inanimate things and living beings. Without going into its details, we can state that living beings obey two kinds of cause-effect relations –*physical laws* and *biological laws*. The former set of laws is normally covered by physics and chemistry. The philosophical aspects of such physical laws have been explained in my previous book *Origin of Universe -- The Light of System Philosophy*. In contrast, the present book is primarily concerned with the second set, designated as biological laws, complied by the interconnected parts of a living being. Hence, it is necessary to start by clarifying the distinction between physical laws and biological laws in a tentative manner.

To illustrate the foregoing point, let us consider the *digestion* happening in a living being. During the process of digestion, food materials are mixed with saliva, acids and enzymes. Due to subsequent chemical reactions, the food items are broken into the molecules of sugars, proteins, fats and amino acids, which are absorbed by the bloodstream. Indigestible substances are excreted. Remarkably, the chemical reactions under digestion would follow the physical laws as mentioned above. However, the various events pertaining to digestion happen through the coordinated function of various organs of digestive system, such as mouth, tongue, intestines and glands. The functions of such organs display specific purposes, and are coordinated according to sequences of time. The notion of *cause-effect* is applied here. We can find general cause-effect relations between the parts of a living being through scientific method; these are called *biological laws*, which differ fundamentally from physical laws.

Let us concentrate now on the various disciplines that are generally called the *biological sciences* consisting of biological laws. For a systematic discussion about these subjects, we must first classify the numerous biological activities of a living being into two separate levels, which are respectively denoted by the concepts *life* and *mind*. These basic terms together indicate the defining characteristics that separate animate world from physical world. The area of scientific knowledge about life is known as biology. There are two stages in its development – classical biology and modern biology. On the other hand, the experimental study of mind comes under the broad area of psychology.

As noted earlier, philosophy of science normally pertains to the area of physical sciences. But some philosophers and thinkers have attempted to extend the philosophical enquiry into the realm of biological sciences also; but are caught in the web of controversies. In this context, the most important debates are with regard to the following questions:

- ❖ Can the biological laws be analyzed through the scheme of received philosophy of science?
- ❖ What is the epistemology (theory of knowledge) of biology and psychology?
- ❖ What is the explanation for the origin of life and mind with regard to the living beings upon earth? Does the religious view about soul and God have any relevance in this context?
- ❖ Is it justified to say scientifically that the evolution of inanimate molecules caused the emergence of life and mind?

Since these questions are still outstanding, without satisfactory answers, we must admit that philosophy of biological science is an unchartered area. The major purpose of present book is to investigate the above issues and to suggest an innovative thesis. Hence this volume is appropriately titled as **Life and Mind: The Light of System Philosophy**. It is the second book in the series of three books dealing with the fundamental aspects of world and phenomena.

System Philosophy is defined here as an integrative thought about the universe as a system of matter and consciousness, where these constituents are in dialectical and productive relation. It effectively shows that things exist by the union of opposites. This philosophical path is to be distinguished from the empirical notion called *systems philosophy* or *systems view*, popularized in the writings of a group of thinkers including mainly David Bohm, Fritjof Capra and Ervin Laszlo.

The unique features of the present book can be picked up by comparing the text with the relevant sections of books used for reference as included in bibliography. All chapters have a considerable set of original ideas; these are clearly marked by the symbol [*]. The main examples of **innovative ideas** are listed below chapter wise.

Chapter 1: *Key Ideas from Previous Book titled "Origin of Universe".* This chapter gives the summary of previous book, which deals with the scientific knowledge about physical (inanimate) world. The controversies and drawbacks of received Philosophy of Science are explained concisely so as to facilitate the introduction of System model of physical world. We have to utilize the findings of this chapter for developing the theory of knowledge about biological world.

Chapter 2 : *Existence of Life.* Firstly, the puzzle of life is explained in terms of different worldviews. The linking of central dogma of molecular biology and genetics to the *machine-algorithm model* is my original idea. Then the epistemological criticism of computer model of life is presented systematically. Treating genetic code (information) as non-physical is the vital step to reach the System Philosophy of Life, to be introduced in this chapter. Here the key idea is that Life is a **system** formed by the opposite entities called macromolecule (mainly DNA) and information. The new System Model of Life synthesizes the religious and materialist ways of describing the origin of life upon earth. It is the first time that a mathematical method – the X-Y diagram of analytical geometry – is adapted to explain philosophical aspects of life.

Chapter 3 : *System Philosophy of Evolution.* This chapter tries to solve the philosophical issues linked to the theory of evolution advanced by Charles Darwin. The foregoing treatise about the origin of life can be extended to the study of biological evolution and formulate its *System Model*. The formation of a new species from an ancestral species happens on account of four factors namely mutation, struggle for life, inheritance and natural selection. *System Philosophy of Biological Evolution* rejects the materialist approach, by holding that each of the four factors is a combination of matter and consciousness. Further we can show that Darwin's theory is a physical reduction of the system model being proposed here. It would equip us to tackle the *problem of missing links* in an ingenious way.

Chapter 4 : *System Philosophy of Mind.* The central objective of this chapter is to refute the doctrine of neuroscientists that mind exists above the *brain and other parts of nervous system (BNS)*. For that purpose, separate Tables are given respectively about the structure of cognitive mind and about the *five factual worldviews* about mind. Neuroscientists do not recognize the nonphysical aspects of mind like creativity, purpose and freedom. It is established here that **human being is a system of matter and consciousness**; this system has three levels of organizations, namely inanimate body, biological body and mind. Human mind has a structure with two main cognitive parts namely, intellectual mind and mystical mind. Consequently, the age-old problem of mind-body dualism is solved through the system model of mind. Also the self-consciousness of human being is explained.

Chapter 5 : *System Philosophy of God and Evil.* We can note that the religious view about soul and God was resorted traditionally as explanation for the origin of life and mind. It is challenged by the advancement of science. We need a proper perspective about the concepts of soul and God; this is the purpose of this chapter. The underlying premise is that religious knowledge is an aspect of human mind, which is in contrast to the faculty for science. The key ideas for originally developing *system philosophy of religion* here are:

- Table of religious beliefs under mystical mind, modifying the theory of Plato
- Definition of *religious life system* (RLS) as the global social system.
- The *methodology, source and justification* of religious propositions.
- The conclusion that the existence of God is an inference confirmed by the mystical experiences of a worshipper.
- The problem of metaphysical realism is removed.
- The application of X-Y model would explain the structure of religious knowledge and also the concepts of God and Evil.

Chapter 6 : *Synthesis of science and religion.* In the first section the central arguments of atheism are firmly refuted. Subsequently, System Philosophy shows the synthesis of science and religion by treating them as two parallel systems of knowledge, which are two levels in terms methodology, source, justification and truth. More specifically, the unification of science and religion is achieved because these are two kinds of knowledge formed by two separate faculties – scientific mind and mystic mind respectively -- of human mind. In the layered view of universe, the Ultimate Reality or *paramporul* is an X-Y coordinate system of body and consciousness. It causes the formation of a hierarchy of things – including inanimate and living things – in the universe. The existence of human mind with various levels of faculties is explained in this view.

As indicated above, the present book presents original ideas about the System Model of life, evolution and mind. Consequently, it is competent to develop a new *philosophy of religion* suitable for the synthesis of science and religion. This book is the result of my self-study and critical thinking of philosophy, spanning over two decades. My acknowledgments to the forerunners of philosophy, science and other kinds of knowledge are briefly recorded with in the forthcoming *Prologue.*

George Luke

Prologue

MY EXPERIMENTS WITH PHILOSOPHY

My life story has many unexpected twists and turns, which reveals the specific intentions of *almighty, the reality of universe*. This brief autobiographical note is intended to reveal the trials and tribulations, which I experienced in the course of my philosophical quest. I was born on 03 June 1953 as the eldest son of Aleykutty and P. L. George belonging to Puthankulam family of Thodupuzha Taluk in Kerala state, south India. Father was a teacher of Malayalam language. I am a member of Roman Catholic Church of Christianity.

Starting life in an underdeveloped part of Kerala state, I studied in Malayalam medium schools - at Neyyassery (1958–62) up to fourth standard and afterwards at Kodikulam (1962-68). In the primary and middle levels I was very poor in mathematics. But during the year of eighth standard I had a sudden interest in Geometry, particularly in proving theorems. It kindled my aptitude for mathematics, which in coming years helped to expand my brain power. I passed SSLC in 1968 meritoriously, having placed at fiftieth rank in the all-Kerala list. I received the national merit scholarship also.

During the vacation after SSLC examination, I started reading books available in local library. Having good proficiency in reading and writing essays, I curiously read books on popular science including the articles that appeared in periodicals like Mathrubhumi weekly, especially relating to Quantum Physics and Cosmology. This interest in scientific topics prompted me to think on the deeper aspects of world and to brood over philosophical questions. My mathematical and logical mind became suitable for creative thinking on the ultimate issues, though I did not have formal education in philosophy.

I Graduated from Newman College (1968-73) at Thodupuzha, securing second rank in B.Sc. (Mathematics Main) of 1973 Batch of Kerala University. It is worth mentioning that I got hundred percent marks in all Mathematics papers. My postgraduate education was in the Department of Statistics, Kariavattam Campus, University of Kerala, Thiruvananthapuram. I secured first rank in M.Sc (Statistics) of 1975 Batch. Though there was an offer of lectureship from a good college, I was interested in higher studies; hence continued in the same Department for newly started M.Phil (Statistics) course and came out in next year with A Grade.

By that time, having studied econometrics in the post graduation level, I was pondering to do research in some practical problem related economics. Keeping this aim in view and due to other circumstances, I joined Reserve Bank of India in Mumbai on 28-09-1976 as Statistical Assistant and became Staff Officer on 28-04-1982. Then I switched over to National Bank for Agriculture and Rural Development (NABARD) and served as Assistant Manager at its Mumbai Head Office during 1983-85. With next promotion to the post of Manager, I was transferred to the regional office at Thiruvananthapuram (1985-1993) and later to Pune (1993-2001). During these twenty-five years, I could acquire professional experience by conducting various studies in connection with official work including inspection of cooperative banks.

While working in Mumbai, my intellectual life was immersed in banking and other fields of economics. I tried to do part-time research in economic department of the University of Mumbai; but my request was turned down due to lack of M. A. degree in economics. This prompted me to conduct self-study of economic texts; but I could not write exam for post graduation in this subject on account of other priorities.

Reaching Thiruvananthapuram, I continued my struggle with serious subjects alongside official work. I can recall that the seed of my interest in philosophy was sown by a special incident in 1988. On a Sunday morning, I was electrocuted while operating the washing machine in my home. Lying in water with electricity flowing through my body, unable to move or speak, I was expecting the imminent death. At that time my co-brother, who was in the next room, came and saw my danger; he rushed to put off the main switch, thus saved my life. This near-death-experience caused churning in my mind and death became a frequent subject for thought. Subsequently, the aptitude for learning philosophy had been growing in the forthcoming years. I used to get shivering due to wonder over the thought that the universe is extended infinitely without end.

Since my temperament was tuned to research activities, the job related to banking appeared to be quite routine and uninspiring. Taking books from the great libraries of British Council and Center for Development Studies in Thiruvananthapuram, I attained considerable depth in various topics of economics. The production function model of classical and neoclassical economics was the center of my intellectual fascination.

During the last period of my stay in the city, my ambition of part-time research in economics was revived. With the guidance of Professor M. A. Oommen, I prepared an article titled *Productivity of Capital Investment on Marine Fishing Crafts of Kerala* and got it published in July-Sept. 1993 issue of *Productivity*, Journal of the National Productivity council, New Delhi. On the basis of this article, I was selected in February, 1993 for research in Dept. of Economics at Kariavattam Campus of the University of Kerala. But the PhD program could not be started due to the immediate transfer of my official posting to Pune, which is reputed for better academic atmosphere. After reaching Pune, I could get the support of a guide for part-time economic research in the Mahe center of Pondicherry University, but I was denied permission on account of the lack of concerned post graduation.

Major highlight of my thought process, in the next phase, is the deviation from the ongoing study of economic themes. As a tryst with destiny, in November 1994, I purchased the popular book of Fritjof Capra, *The Turning Point,* which was originally published in 1982. Here Capra describes various layers of world – subatomic phenomena, biological organisms and different social organizations – as evolving systems, which are wholes of interconnected parts. Then he resorts to Chinese mysticism for addressing the ultimate questions. Since a whole is more than the sum of parts, it has holistic, ecological and dynamic existence. Accordingly, Capra's *systems view* of life and economy appeared to counter the mechanistic approach of the corresponding disciplines.

The treatment of systems in Capra's book triggered my critical mind and it became the turning point in my intellectual pursuit. I recognized that Capra has not explained the origin of life in the inanimate macromolecules like DNA. And, he failed to present a philosophical perspective about systems. What is the philosophy about the interconnected but layered world? How can we compare mysticism with the alternative thoughts about reality? What are the essential principles behind capitalism and communism? *These questions agitated me and I decided to take up the study of philosophy as a new venture.* I enlarged my reading by purchasing hundreds of useful books in philosophy, social sciences and related subjects from exhibitions and book stalls.

The persistence of poverty and high inequality of income-asset distribution had been a matter of grievous concern to me, in the course of analyzing the economic problems. Another issue that pained me was the cruelty and destruction due to wars as well as terrorism. Why do political leaders commit such atrocities, even though religions profess love and peace? Pondering over these problems I came to the conclusion that most of the crimes and evils are performed at the social level, rather than at individual level. The main drawback of idealism and religious philosophy is its focus on individual mind without giving due importance to the patterns of social behavior.

Thus I began to use the social perspective for deliberating about world. The most important challenge for me was to explain that social systems exist by the complementary relation of opposites. Matter-energy, space-time, body-mind, self-society, capital-labour are prominent examples of opposites, when we consider various phenomena. Continuous exposure to my books and thinking over different aspects of world generated a unique idea in me that our life is spent in seven global social systems; this principle is denoted by the phrase *seven life systems*. It became the key to start my philosophical project, which was named *System Philosophy*.

In this context, I got an intuitive idea that the *production function model* of economics would give an innovative method to study the interconnected behavior of opposite entities as well as social systems. However, for articulating this proposal, I have to study philosophical doctrines seriously. Gradually I realized that the routine of official job is a hindrance to my progress in this direction. Moreover, I was suffering from diabetes for some years on account of the continuous strain of banking career and private study. Due to the pressure of such circumstances, I voluntarily retired from NABARD service on 29-9-2001 for engaging with the research, writing and publication in the field of philosophy.

Sitting at home, I plunged into the selected books in my possession for developing the themes of System Philosophy. The dialectical method that I resorted was to take notes in English and then translate it into the mother tongue Malayalam. Then the reverse method also was adopted. This bilingual process has helped me to increase the clarity of philosophical issues.

My deep interest in the production function model of economics generated a novel idea in me for depicting the reality of universe as well as our social systems. It is the first time that a mathematical model is employed in philosophy for explaining its abstract concepts.

Inspired by the hope to articulate System Philosophy in an innovative manner, I engaged myself for three years in preparing the manuscript of my first book in Malayalam. Then I approached certain important publishers, but they refused to publish my philosophical book, holding that it is difficult to sell in current scenario. So I self-published the book in April 2004 with the title **Saptaloka Darshanam**

Samgraham *(philosophy of seven life systems – a summary)* under the banner of PGL Books, Changanacherry, Kerala – 686 101. This work was mainly intended to explain the theoretical concepts of the seven global systems namely *nature, economy, politics, family, ethics, religion* and *art*. Due to the popular and traditional importance of number *seven*, many readers were amazed at my classification.

Next development is my participation during 2004-2005 in the program of the School of People's Economics, conducted by the NGO called VICHARA at Mavelikara, Kerala. It consisted of about thirty days of discussions and seminars on various topics, which sharpened my philosophical ideas. As a result, I prepared an article titled *From Modern Science to System Philosophy* and published it in the June 2005 issue of the journal OMEGA of ISR Aluva, Kerala.

As a matter of divine providence, I got the opportunity to join a research program during 2005-2008 under the Association of Science, Society and Religion (ASSR) of Jnanadeep Vidya Peedh (Papel Seminary), Pune. The discussions and seminars conducted in December month of these years as well as the library facilities helped me to research on the interface between science, religion and philosophy. I presented a dissertation on this topic and it became the spring board for my intensive pursuit in the following years. So far I have purchased over thousand academic books that serve as authentic references for full time creative work, while remaining in the modest facilities of home.

- In March 2015, I self-published an authentic book ***jeevanum parinamavum*** (Life and Evolution) in Malayalam language discussing the theories of biological phenomena in the light of *System Philosophy*.
- I got an article titled *Ayurvedathinte Jnana Sithantham* (Theory of Knowledge of Ayurveda) published in the May &June 2017 issues of OUSHADHAM Journal of Ayurvedic Medicine Manufacturers Organization of India (AMMOI), Thrissur, Kerala.
- Organized the *Academy of System Philosophy* under a trust to disseminate the new philosophy among wider audience.

The foregoing is the background for preparing the manuscripts of the comprehensive books on *System Philosophy*. In the recent years I have exchanged ideas with many well placed scholars; but to my surprise they are finding it extremely hard to understand the principle of system as elaborated in my writings. Generally people are obsessed by the thought that opposite entities are separate, without any interconnection. Here I may mention the fact that acquiring knowledge is a social process, which is influenced by the ideologies and vested interests of powerful individuals of society. If a new idea comes from a lover of wisdom, who lacks the support of institutions like universities and media, it will normally face the struggle for existence. I believe that the esteemed readers of my book will help its natural selection in future because human mind has an innate tendency to prefer truth and discard falsehood.

Many teachers, friends, relatives and well wishers have helped me in the course of my life so as to contribute to the evolution of my philosophic views. My deep gratitude to all of them is beyond words. I would specifically thank Professor Hardev Singh Virk, for showing keen interest in reading my previous articles, perusing the manuscript and finally gifting me with an introduction to the present book.

I am especially indebted to many writers of philosophy and related subjects, as mentioned in the notes of chapters as well as the bibliography of this volume. I can emphasize that my limited words are not sufficient to express my acknowledgement of the ideas received from the forerunners in the history of thought.

George Luke
03 – 06 - 2018

Introduction

Life and Mind : *The Light of System Philosophy* is the second book in the series of three books dealing with the fundamental aspects of world and phenomena. The other volumes are respectively *Origin of Universe* and *Discovery of Reality*. As a combination it forms a comprehensive treatise on REALITY by George Luke who took voluntary retirement at the age of forty eight to prepare his *magnum opus*. The author has distinguished the phrase *system philosophy* from the alternative concept of *systems philosophy*, which has been popularized in the writings of Ervin Laszlo. In contrast, George Luke defines *system philosophy* as an integrative thought about the universe as a system of matter and consciousness, where these constituents are in dialectical and productive relation. This philosophical perspective about real existence involves the synthesis of rational and empirical aspects of knowledge.

The present book has been divided into seven chapters.

Chapter 1 deals with a summary of the treatise of the first book *Origin of Universe*. The important contents are the following:

- Fundamental Ideas of Physical Science
- Big Bang Cosmology and Quantum Cosmology
- Philosophy of Science
- System Philosophy of Science
- Cosmological Puzzles Finally Solved

In a summary way we can say that the dilemma about matter and energy has been highlighted focusing on the question of existence and failures of scientific realism. The author resorts to his system model, which gives a practical and phenomenal theory about the existence of physical world.

The second chapter of present volume is focused on Life, firstly giving account of various theories about life, mainly genetics, and its critique. Here the author defines System in a philosophical sense for showing the existence of phenomena and reality. This principle is then used for developing the system philosophy of life.

In the **third chapter**, the author refers to pitfalls of theories of evolution and writes as follows: "But many theologians and religious fundamentalists opposed Darwin's theory and evolutionary science accusing it as challenging the belief in God. Accordingly, they have published great volume of literature supporting the biblical story of creation. This aggravated the conflict between science and religion".

In the opening paragraphs of **Chapter 4** "System Philosophy of Mind", the author writes: "The totality of mental activities is conventionally termed as **mind**. Accordingly, we hold that ***mind is a higher phenomenon which exists over and above the biological processes of nervous system.*** The foremost issue in *philosophy of mind* is the definition of mind since we have to take into account the related notions like body, soul, spirit and consciousness. There are great differences between science and religion while considering the question: what is mind?" Then the author proceeds to develop the system model of human mind.

In Chapter 5, "System Philosophy of God and Evil" author refers to anatomy of a crisis in the Christian world. I appreciate the idea of author to define religion as a social system: "It is a popular notion that *religion is a social system* in view of the empirical and concrete aspects like various activities of worship, the organizational structure of churches and temples as well as the social relations between the believers".

Finally, **Chapter 6** discusses the conflict between the assertions of science and religion. The outstanding problem is introduced by the author as following: Science denies the existence of supernatural beings or forces. It reduces the natural things, which are actually composed of body and mind, into forms of matter and energy. That is, science holds that everything in the nature is *physical*. It formulates cause-effect relations, known as *physical laws*, based on sensory or experimental evidences about the properties of physical things through our *scientific mind*. This is public knowledge in third person perspective. On the other hand, religion presumes that supernatural beings or forces exist and they involve in the affairs of natural world. This is based on mystic experiences through revelation, emotions, ecstasy and meditation of religious leaders as well as ordinary believers. The mystic ideas are private and beyond sensory experience - it is the function of *religious mind*.

Happily, a coherent argument is developed for solving the issue. I may quote from the end part of author's thesis: *"System Philosophy shows the synthesis of science and religion by treating them as two parallel systems of knowledge, which are two levels in terms methodology, source, justification and truth.* More specifically, the unification of science and religion is achieved because these are two kinds of knowledge formed by two separate faculties – scientific mind and mystic mind respectively -- of human mind. In the layered view of universe, the Ultimate Reality or *paramporul* is an X-Y coordinate system of body and consciousness. It causes the formation of a hierarchy of things – including inanimate and living things – in the universe. The existence of human mind with various levels of faculties is explained in this view".

As per the foregoing, the present book presents original theses about existence of life and biological evolution as well as the philosophy of mind. Further treatises are philosophy of religion as well as science-religion synthesis.

George Luke has done a commendable job in preparing three volumes - Origin *of Universe, Life and Mind* and *Discovery of Reality* - using concepts of System Philosophy. I congratulate him for this singular achievement.

Hardev Singh Virk
Professor of Eminence
Punjabi University
Patiala, Punjab (India)
www.drhsvirk.weebly.com

14 February, 2018

Chapter 1

Key Ideas from Previous Book

****

1.1 The Framework of Physical Science

1.2 Dilemmas about the Origin of Universe

1.3 Philosophy of Science and its Failure

1.4 Vision of System Philosophy of Science

Author's main original ideas are marked by [*].

The mark [#] gives the number of note at the end.

****

An expedient summary of the previous book *Origin of Universe - The Light of System Philosophy* is given in this chapter to prepare the ground for further deliberation of the philosophical issues regarding natural world. The said book is a thesis about physical world and it contains a good number of original ideas. [# 1]

1.1 The Framework of Physical Science

The framework for scientific enterprise consists of the foundational ideas such as matter, energy, space and time, which together constitute the notion of physical world. In other words, the assumption about the existence and properties of matter forms the basis for the whole edifice of science

An important philosophical concept called **worldview** is used for understanding different paradigms of physical science and other areas of knowledge. Here we define **worldview** as the set of common basic ideas found in the broadest family of theories. In other words, it contains the essential ideas of all theories having family resemblance. Hence, worldview is the method for classifying the totality of theories into the largest groups. Thus there are **six worldviews** adopted in the history of human thought, which can be arranged as below. [*]

1. Organic worldview (OWV)
2. Mechanistic worldview-rational (MWV-R)
3. Mechanistic worldview-empirical (MWV-E)
4. Spiritual process worldview- rational (SPWV-R)
5. Spiritual process worldview-empirical (SPWV-E)
6. Physical process worldview-empirical (PPWV)

The classification of six worldviews enables us to systematically draw the landscape of numerous theories pertaining to value and fact in diverse fields of knowledge. Also it is essential for comparing and contrasting the important doctrines of pioneering philosophers.

From our commonsense as well as scientific points of view, there are **three levels of universe** that can be experienced through our sense organs: inanimate world, animate world and social world. Obviously we exclude the metaphysical concepts like God, gods, angels and evils from our purview. In accordance with the main divisions of universe as above, different branches of science have been developed focusing on separate areas of study. Generally speaking, science is the systematic study of natural things using experiments and analysis of data in order to determine the *cause-effect relations*.

The foundation of physical science is the principle about the existence of matter including various forms of energy. Scientist derives this idea from the *mechanistic worldview* and the *physical process worldview* introduced above. These worldviews respectively are the foundations of the two stages in the development of physical science -- Classical Science and Modern Science. Further, it can be shown below that these stages essentially pertain to different ways of conceiving the characteristics of matter and energy.

The paradigm of physics underwent a radical change from Classical Science to Modern Science in the early decades of 20th century on account of the discovery that atom has component particles; thus it rejected the earlier theory that atom is indivisible. By 1930, the planetary model of atom was accepted generally. Four basic forces determine the planetary model of atoms as well as the various processes within atoms and higher substances. Atom has *four subatomic particles*, namely proton, neutron, electron and neutrino as well as *four basic forces*, namely gravitational force, electromagnetic force, strong nuclear force and weak nuclear force.

The knowledge about the properties of the above subatomic phenomena as well as their mutual relations would constitute the modern branch of physics called **quantum mechanics**; it revolutionized our knowledge about the nature of physical world made of matter and energy. The term 'quantum physics' refers to the wider discipline including the areas of quantum mechanics, quantum field theory, astronomy, cosmology and related enquiries.

The principle of particle-wave duality prompted theoretical physicists to abandon mechanistic world view for studying the characteristics of subatomic particles and forces. Alternatively, they proposed the **physical process worldview** as the new paradigm for finding the laws about subatomic phenomena with particle-wave duality. It envisages describing the activities of such phenomena, without considering the issue of existence of particular forms of matter and energy. We may emphasize here that the discipline of *quantum mechanics* was developed on the basis of physical process worldview. Accordingly, we can say that quantum mechanics adopts the **machine-algorithm model**. The subatomic particles with energy exist like a machine part and the movement (activity) is like an *algorithm*.

The laws of quantum mechanics are capable of explaining the physical properties and chemical activities of substances belonging to the macro level. Taking into account these facts we may postulate that the term *physical world* refers to the objects of classical science and quantum mechanics.

Subsequently, quantum field theory (QFT) was proposed for achieving the unification of subatomic particles and basic forces. But the experimental and theoretical research in this direction resulted in the discovery of more elementary components, which are neatly arranged in a scheme called **Standard Model**. Its most simple and elegant form consists of 18 kinds of fundamental particles -- 12 kinds of material particles (quarks and leptons) and six kinds of energy particles (bosons). But, the elaborate picture of standard model consists of 61 kinds of fundamental particles. The principles of quantum field theory and **Standard Model** are applied for explaining the fundamental constitution of matter. This naturally leads us to the deliberation about the origin and evolution of material world. For explaining the origin of matter and the formation of physical world, scientists resort to the subject of *cosmology*. Obviously it is the most basic discipline of physical science; the concerned outstanding issues will be discussed in the following section.

1.2 Dilemmas about the Origin of Universe

We can define cosmology as the scientific study of the past history of universe, treating it as physical world made of matter. Taking into account the various theories of cosmology proposed by scientists, we note that the most important are *Big Bang Cosmology* and *Quantum Cosmology.*

We can find that big bang cosmology is insufficient. Quantum cosmology tries to unify the diverse components of standard model for reaching at an ultimate cause of physical world. Now quantum cosmology is challenged by important **puzzles**, which are summarized in the following serious **drawbacks**. [*]

- ❖ The search for the most elementary form of matter is elusive, because physicists are increasingly resorting to mathematical models associated with the notions of symmetry and group theory. The experimental evidence is lacking, or at least ambiguous, in the case of the said mathematical models.
- ❖ Consequently, we can reasonably doubt whether the fundamental stages of matter have real existence. Our commonplace notion of matter as an extended substance cannot be supported by cosmological theories.
- ❖ Scientists have failed in explaining the evolution of matter to more complicated forms and also the emergence life and mind causing various kinds of biological cells and organisms. The biological world abounds in nonphysical aspects of creativity and purpose, which are outside the scope of science.

For making head way in this difficult situation, it is necessary to analyze the scientific knowledge in general and cosmology in particular from philosophical perspective. So we are now concerned with **philosophy of science**, which can be alternatively designated as the epistemology (theory of knowledge) of science. We will see in due course that the epistemological aspects of the cosmological doctrines constitute the deeper levels of philosophy of science. [# 2]

1.3 Philosophy of Science and its Failures

The scientific laws or cause-effect relations are generally obtained through the sequence of theory (Ty), hypothesis (H), deduction (D), testing (T) and inductive inference (I). These stages are ordered like the organs of an animal. The propositions coming under successive stages of Ty, H and D together is called *deductive propositions (DP)*; the propositions of T and I are collectively designated as *inductive propositions (IP)*. We have suggested the new phrase *TyHDTI scheme* to denote the scientific method of combining DP and IP to produce scientific laws.[*]

Though we ordinarily say that scientific laws are inferred from repeated experiments, philosophical investigation would reveal the role of preconceived abstract ideas in the experimental process. So we originally propose that each scientific law is the result of five stages of TyHDTI scheme. The aim of epistemology is to determine the validity and truth of such scientific laws. We innovatively propose that the vital components of epistemology are **methodology, source, justification and *truth*.** [*]

Then the connotations of the four components of epistemology are given below:

- ***Methodology*** is the deliberation about the general components – Ty, H, D, T and I -- of scientific method as well as about the relative importance of DP and IP in the meaning of scientific laws. Also, the various definitions and meanings given to the fundamental terms are clarified in different theoretical situations.
- ***Source*** denotes the theory about the structure of scientific mind, which generates the diverse kind of propositions under DP and IP.
- ***Justification*** deals with the issue whether the scientific law represents actually existing aspects of universe. Accordingly, we must get sufficient evidences to judge that the concerned scientific law is valid. *The essence of justification is that the components of theory -- namely space, time, matter, energy, and so on – are not imaginary concepts, but are the fundamental aspects of physical world.* It finally leads to the question: does matter or physical world exist?
- ***Truth*** is the quality of a justified belief under the TyHDTI scheme, when it conforms to the actually existing things of universe. Moreover, truth is the unifying principle applicable to all kinds of knowledge such as science, religion and art.

We know that **classical science** adopts *content view*, implying that the concerned propositions represent actual objects. It is in tune with the principles of mechanistic worldview. Consequently, the propositions under DP and IP are not unified, since these are treated to have separate kinds of validly and meaning. Then there are two opposite doctrines – **rationalism** and **empiricism** -- about *methodology* and *source* of the physical laws. But we can state that these opposite positions cannot be reconciled within the framework of mechanistic worldview.

In the case of **quantum mechanics**, the subatomic phenomena have the wonderful property of particle-wave duality, necessitating the physical process view. Instead of treating subatomic particles and forces as separate entities, quantum mechanics focuses on the activities of phenomena involving matter and energy. The concerned methodological doctrine is called *logical positivism*. It is alternatively known as *the verifiability criterion of meaning* or *the verification principle*. It also implies that a word will not have meaning if there is no evidence to verify it. According to this line of thinking, logical positivists decree that the religious words like God, Soul, spirit, heaven and so on are meaningless.

The issue of reconciling rationalism and empiricism in respect of the diverse propositions under DP and IP would come up in the case of quantum mechanics also.

In a ground-breaking manner, we propose that the methodology of *quantum cosmology* is Modern Phenomenalism (simply, Mophism) because logical positivism fails here miserably. Mophism strictly follows the agenda of materialism and empiricism. *The methodological dilemma in this stage is similar to that of classical science because the tenets of Mophism can be challenged by the doctrine of rationalism.* Regarding the topic of *source* pertaining to the models of quantum cosmology, scientists adopt the empirical philosophy of mind called *computer model functionalism*, which has serious drawbacks.

Now we will take up the problem of **justification** with respect to the three paradigms - classical science, quantum mechanics and quantum cosmology. As mentioned earlier, *justification* means the validation of a law on the basis of sufficient evidences. It involves the production of arguments to prove that the state of affairs represented by the law exists actually.

The theory of justification under classical science is called ***naïve realism***. It leads to skepticism about the origin of matter. But scientists choose to continue with their empirical studies without considering its philosophical drawbacks.

The problem of justification pertaining to quantum cosmology is still worse. We may naively think that cosmology has validity only if the cosmological entities have real existence. But, how can one say that the concerned mathematical models represent real aspects of the early stage of universe? Physicists adopted the position of **scientific realism,** which attributes real existence to theoretical entities of cosmology from practical point of view. But it suffers from the worst form of skepticism. Further recalling the drawbacks, given earlier, of the recent proposals of quantum cosmology, we have to reject scientific realism.

1.4 Vision of System Philosophy of Science

Aristotle originally proposed that everybody must apply three rules of thought for knowing the individual things separately. It means that we can think about things – particles or metaphysical beings – distinctly as parts of universe. Moreover, we assume in a common sense way that such things have separate existence.

The fundamental tenet of System Philosophy is the definition of **system** as a productive structure composed of opposite entities. The opposite components of a system are similar to the X axis and Y axis pertaining to the coordinate model of analytical geometry. This system model is illustrated by a diagram to be shown in the next chapter of the present book *Life and Mind*. Hence, system is a symmetrical structure of X-Y coordinates. Here X and Y have dialectical and productive relation similar to that of a factory. *As a rejection of Aristotle's rules of thought,* the opposite entities X and Y are not independent entities. Instead, X and Y are opposites having complementary character. We cannot define one entity without considering its opposite entity. The opposite components of a system are complementary to each other; they have interdependent existence. [*]

There are many main levels of theories for physical science to produce knowledge through TyHDTI schemes – these are conveniently classified into quantum cosmology, quantum mechanics and classical science. We must effectively unify the propositions under deductive propositions (Ty, H, D) and inductive propositions (T, I). Such a synthesis would come from the *system philosophy of mind*. In anticipation of that doctrine, the relevant points are presented in the following paragraphs.

In the light of system philosophy of mind, we propose that human mind is a system formed by the dual aspects of *brain and other parts of nervous system (BNS)* and *consciousness*. Mind has many levels, where each is a system with X-Y model, where X and Y denote the concerned parts of BNS and consciousness respectively. So we can conceive scientific mind like a factory for producing two classes of propositions denoted as DP and IP. Now we know that DP and IP are two levels of propositions where both have the aspects of X and Y as indicated above. In this situation, DP and IP are complementary opposites which constitute a whole of knowledge as per the TyHDTI scheme. [*]

It is possible to hold that the problem of induction is partly removed by using a **good theory** as per the TyHDTI scheme. In other words, a good theory would save a scientific law from the problem of induction up to a reasonable extent. However, there is still an element of uncertainty about the repetition of scientific law in future instances because future is unknowable per se. Considering the role of a good theory, we can suggest that there is not much significance to Hume's problem of induction.[*].

The problem of justification hinges on the questions: Does physical world (matter) exist? What is the origin of matter? My answer to the basic cosmological question is given by the *System Model of physical world*. It describes the universe as a system of matter and energy.

The past universe (parental universe) is a system of dark matter and dark energy, which are denoted by DM and DE respectively. **This system has existence** since it is represented by the X-Y model of coordinate geometry. Then big bang is the origin (x = 0, y = 0) of the system model. The opposite entities DM and DE can undergo the process of compactification throughout the history of present universe; it accounts for the increase of matter from zero to 10^{50} tons.

In order to show the existence of past and present universe in a historical manner, we can use similar X-Y model. Then, the past universe appears in third quadrant while the present universe (physical world) belongs to the first quadrant. The X-Y coordinate system is called the **System Model of physical world** according to the content view. Since the model can be transformed into another X-Y model, where X denotes space and Y denotes time, we can answer the questions about the event of Big Bang. It is emphasized that *origin* and *evolution* are physical notions, which are to be conceived in the framework of space and time.

When we ordinarily say that our physical world originated through big bang, we assume that big bang actually happened. It is the position of scientific realism which must be refuted. For that purpose, we hold that the above *system model of physical world* is a theory. Then the TyHDTI scheme is applied to get the inference that physical universe exists. In other words, **existence is an inference** produced by the cognitive mind of a person; the realism is avoided through this way. Also, the *system model* effectively removes the problem of pluralism.

On further deliberation, we must admit that there is *nonphysical* purpose and creativity in nature. This feature is visible in inanimate world (including galaxies, stars and planets) and more clearly in the phenomena of biological world. It will be shown in due course that nature is a *matter-consciousness system*. We do not require *the intelligent design argument* for explaining the *nonphysical aspects*. Without assuming realism, we can show that the scientific view of physical world is the materialist reduction of matter-consciousness system. More specifically, the matter-energy system is the **physical reduction** of matter-consciousness system of inanimate world. [*]

When we say that physical world (matter-energy system) is the construction of our scientific mind, we mean that it is phenomenal. Since justification is a part of secular epistemology, it has scientific connotation. In the layered view of universe, the topic of justification is concerned with the separate and hierarchical existence of various levels of systems which are broadly grouped into inanimate, biological and mental worlds. Later we will develop the *System Philosophy of Ultimate Reality* showing that the Ultimate Reality or *paramporul* is depicted by X-Y coordinate system of body and consciousness. [*]

The creativity and purpose of physical world is basically explained by the component of consciousness in the foregoing system model. It can effectively solve the conundrum of God and Intelligent Designer in the context of science [*].

Now we consider the common sense question: **what is matter?** It is explained that *the concept of matter does not have absolute and practical relevance in the cosmological stages*. Considering the features of subatomic level and visible level, we adopt the **practical notion of matter**. The property of extension of matter is basically due to the planetary structure of atoms; such atoms constitute the visible structure of physical world. Since the absolute notion of particle is not relevant in subatomic world, then we cannot say that matter exists as a subatomic particle. *Accordingly, the word matter refers to the individual atoms and the higher substances formed by them.* This matter is constituted by subatomic particles and basic forces. Thus we reach at the solution for the puzzle of matter.

Since Intelligent Designer is a jump of imagination from finite evidences to the realm of infinity, there is no coherent meaning for that concept. Unfortunately a section of scientists and theologians have attempted to promote the idea of Intelligent Designer, without knowing its epistemological drawbacks. In a summary way, we can assert that there is no justification for their empirical inference about an infinite being as Designer.

Thus, in the foregoing **System Cosmology**, we have presented the solution of the cosmological puzzles of matter, big bang, space, time, superstrings, membranes, dark matter and dark energy.

NOTES of chapter 1

\# 1   The Selected Bibliography used for the previous book *Origin of Universe - The Light of System Philosophy*, is given below.

Beiser (2002), Bird (2003), Capra (1983), Capra (1992), Davies (1995), Davies (2007), Grayling (Editor) (1995), Green (2005), Gribbin (2008), Hawking (1995), Hawking (2011), Lavin (1989), Martin Curd and J. A. Cover (1998), Michio Kaku and Jennifer Thompson (2007), Newton (2010), Robert John Russell (Editor) (2004), Rosenberg (2000), Smolin (2008), Tarnas (1991), Urmson J.O. and Jonathan Ree (1989).

\# 2   The topic of Philosophy of Science will be elaborated further in the ensuing chapters of the present volume *Life and Mind*.

\# 3   Here the references are: Hawking (1995) pages 149 and 185; Hawking (2011) pages 38-42 and 207-210.

Chapter 2

Existence of Life

****

2.1 The Puzzle of Life

2.2 Modern Biology

2.3 Criticism on Computer Model of Life

2.4 System Philosophy of Life

    2.4.1 Philosophical Definition of System

    2.4.2 Cell is a Phenomenal System

Author's main original ideas are marked by [*]

The mark [#] gives the number of note at the end.

****

A thorough exposition of the philosophy of physical science, as given in the previous book, would naturally lead us to the consideration of knowledge about biological world. Recent three centuries witnessed the development of Biology adopting the physical approach. But the living beings have certain distinct features, which deserve our attention. Here the crucial question is whether the laws of physical science – mainly physics and chemistry -- can be applied to the cause-effect relations observable in living world. A philosophical enquiry into this issue is the chief objective of this chapter.

The origin of life and its existence on the earth has become an interesting topic in last 200 years because of the discovery that a living being can originate from other living beings only. Combining this point with the notion of evolution, scientists began to speculate about the hierarchy of living beings starting from the first occurrence of minute organisms about 3.5 billion years ago. It means that the origin of life is consequent to the evolution of material things during the earlier stage of earth's history. The emergence of life is an epoch that needs explanation from scientific and philosophical point of view. The secret of life will be revealed only if we answer the following questions: What is life? How do we get knowledge about life? What is the theory of modern biology? How did life originate in the earliest form of living beings called bacteria? What is the relation between life and material body? How can we explain the biological evolution leading to the diversity and hierarchy of living beings?

The ensuing sections aim to expose the lacunae in the theories proposed hitherto regarding the questions given above. A proper understanding of the fundamental ideas of classical biology and modern biology – with the main branches called cell biology, molecular biology and genetics – is required for a meaningful discussion about life and evolution. We will focus on the epistemology of modern biology to see whether its inferences have validity. By criticizing the concerned biological theories we will proceed to develop the System Philosophy of Life and Evolution. This innovative philosophical framework will radically interpret the latest scientific achievements in biology and mark its limitations.[# 1].

2.1 The Puzzle of Life

The common division of things into two classes – namely, inanimate things and living beings – is based on our scientific consciousness. The individual beings in the living world are called *organisms*. The building block of an organism is **cell**. We can compare cells to the bricks used for building a house. Since cell is the basic structure having life it is practically important to divide cell into body and life. Body is the material structure of the cell and it is composed of certain inanimate molecules. But the material molecules are organized to form a cell just because of the presence of life. In other words, we can separately consider the body and life pertaining to a cell and it is the context for raising the question: What is life?

Our earth is the home for millions of types of organisms which range from unicellular organisms like bacteria to complex organisms with billions of cells such as plants, fishes, birds, reptiles, animals and humans. In the organisms having more than one cell, body is the material structure formed by the cells. In higher organisms, especially humans, every individual has three levels – *body, life and mind*. Here, body is the lifeless material structure; we may equate it to the dead body. But it is a common practice to use the word 'body' to refer to the biological body (the organization of cells having life), and it causes some confusion. So we will stick to the definition of body as the inanimate material structure of an organism. Also, it must be borne in mind that the activity of life is required for the growth and organization of cells, so that a particular bodily structure is formed. The complimentary existence of body and life makes the topic of life a puzzle.

The difference in the size, shape and other features of cells is due to the *specialization* of cells for different purposes. A particular *tissue* is formed by arranging the cells of a given kind in a specific manner. There are about 200 kinds of tissues in human body. You can note that the layers of skin, the fibers of muscles, tubular structure of intestines, hard nature of bones and the liquid state of blood are examples of specific arrangement of cells with concerned features. In the next stage separate groups of tissues form the organs, which are meant to perform certain biological activities of the organism. Finally we can see different organ systems, each consisting of a group of organs; the blood circulation system, digestive system and nervous system are quick examples.

Next we shall consider the aspect of mind, which is very evident in a human being. Now it is necessary to divide our biological body into mechanical organs and nervous system. The activities such as breathing, metabolism and blood circulation are performed by mechanical organs. The nervous system consists of brain and numerous nerves spread all over the body; it is made of a special class of cells called neurons. The biological activities of nervous system result in the formation of networks of neurons. Consequently, the phenomenon of mind occurs as composed of ideas, emotions, feelings, memory etc. Thus, mind is a higher level of function supervening the neuron network. There is evidence that most of the animals also have mind, though at simpler levels as compared to human beings. *But it is conventional to hold that mind exists only in those living beings which have brains and nerves.* Accordingly, the lowest group of animals as well as the entire species of plants do not have mind. A practical distinction between life and mind is necessary for discussing the philosophical issues pertaining to biological world.

The main biological activities or the characteristics of life are listed below in a simple manner:

1. *Metabolism:* It is the activity of digestion in which food items are processed for generating energy, absorbing useful elements and finally excreting waste materials.
2. *Growth:* It is important to note that life of an organism starts from a single cell at the earliest stage of an embryo. Hence, the growth of organism occurs due to the increase of cells caused by the division of pre-existing cells and subsequent duplication. The cells in a biological body are continuously decaying and are replaced by new ones. The net effect of decay and replacement of cells determine the stages of growth, maturity and aging of the organism. As a result of aging, the organism finally dies. Every organism has a particular lifespan between birth and death.
3. *Adaptation*: Organism is able to react with and adapt to external circumstances. Accordingly, it causes the movement of its bodily parts. In certain cases the body of an organism

undergoes variation for adapting to the environment. When this adaptive process is continued through many generations it may lead to evolution, that is, the origin of new species.
4. *Reproduction*: An organism originates from previously existing organism; this process is called reproduction. In the case of higher organisms, the parents (male and female) participate to produce offspring in the form of a cell.
5. *Self-maintenance*: The cell is capable of maintaining itself. It involves the repair of damaged cells and at the same time the production of new cells and combining them to form tissues and organs.

It is interesting to ask whether life exists in any other part of universe also. So far scientific evidence confirms that life exists only upon earth. This renders the origin and development of life on earth as a subject of philosophical importance. We cannot confine the study of life to the realm of science, as there are questions about the meaning and purpose of life that is uniquely found in this planet. The enigma of life is primarily due to the fact that religion and science proposed divergent concepts about life.

Different worldviews

As per the received documents, we approach the question "what is life?" through two paths: First is to treat life as a being or substance, which is responsible for the specific biological activities listed above – this is content view. Second path is to hold that life is not a substance, but is an activity or process. Further it is necessary to divide both these paths into rational way and empirical way. According to this method, we can fundamentally classify the various theories adopted into the table of worldviews as illustrated below.

Table 1 : The Worldviews about Life[# 2][*]

	Content view	**Process view**
Rationalism	Organic world view and mechanistic world view-rational *(vitalism)*	Spiritual process view – rational *(romanticism / evolutionary theology)*
Empiricism	Mechanistic world view- empirical *(Naturalism or classical biology)*	a) Spiritual process view – empirical *(prana principle, yoga, etc.)* b) Physical process view *(modern biology / computer model)*

Vitalism is aligned with the religious notion of soul. Accordingly, the birth of a person occurs when immortal Soul enters the body, whereas death is the event of departure of Soul from the body, like a bird flying out of its nest. In this way, life is the activity of soul upon body. So life can be treated as a metaphysical force called *vital force* – this view is known as **Vitalism**. It means that life is the manifestation of a metaphysical being, which is different from material body. The influence of metaphysical realism is so strong in our mind that we have the habit of separating life from physical body. The forms of language create confusion when we say the following typical sentences: "death is the event when life departs from body", "this is a dead body" and "he lost life". It can be anticipated that this religious view involves many philosophical problems.

According to the rational side of spiritual process view, God is the immanent force working in all things of this universe; it manifests as vital force determining the activities of organisms. Following this line of thought, Romanticism appeared as a form of mystical thought in 16th century introducing the idea of biological evolution. In 1809, Lamarck originally presented the first mystical theory of evolution, and it was subsequently elaborated by the philosophers like Teilhard Chardin and Albert North Whitehead in 20th century. According to them, the origin of new species is due to the divine activity working through immanent life force. This mystic view of biological evolution is alternatively called *evolutionary theology*. [# 3].

The empirical side of spiritual process view is prominent in the eastern mysticism of China and India, where the process of life is called *prana*. Various medical practices have been developed on the basis of the *prana principle*. Main examples are *acupuncture, ayurveda, yoga, kundalini and pranic healing*. Another doctrine holds that there is an energy sphere or *aura* surrounding human body. Additionally, there are certain energy centers called *chakras* that control the activities of human being. A separate branch of subject termed *parapsychology* has been developed in recent times aiming to conduct experimental study about paranormal phenomena, which are manifestations of the mystical behaviors of people. But such aspects of life cannot be brought under pure science; hence parapsychology is derided as *pseudoscience*. The multitude of mystical doctrines can be seen as the attempt to describe the enigmatic features of life in metaphoric and imaginative language. [# 4].

Consequent to the Renaissance of 15th century, the scientific study of life started as a counter movement to the above religious approach. Then biological phenomenon is reduced into the physical aspects of body, to be analyzed through the method of physical science. We can note two stages in this physical approach -- classical biology and modern biology. These streams of disciplines are concisely introduced below in the light of concerned worldviews.

George Luke

Historical Development of Biology

The empirical version of *mechanistic world view* formed the basis of **classical biology** as advanced in 18th and 19th centuries. It treats organism as a machine with interconnected parts. Hence, classical biology is generally called *mechanism*. The essential point is that life is not a metaphysical being or vital force, but is a byproduct of the physical and chemical activities happening in various parts of body. Hence, a biological organism is a complex arrangement of material molecules, mainly compounds of carbon, hydrogen, oxygen and nitrogen. Life exists as a separate entity that is an *epiphenomenon* of material structure. In this approach, the scientific knowledge about biological body is derived through sensory observations according to the theory of knowledge called Empiricism. Here, the ontological theory is *naturalism,* which is an extension of materialism to the realm of biological phenomena. So we can reiterate that naturalism together with the empirical method constituted the structure of classical biology. Its main branches are Zoology pertaining to animal kingdom and Botany dealing with plant kingdom.

We can see that the disciplines of classical biology were developed through two main approaches of experimental inferences:

- The cause-effect relations between various organs as well as between the parts of organs were studied by applying the mechanical laws of classical physics.
- The analysis of chemical and electrical reactions happening in various components of biological body was conducted using the laws of chemistry.

For example, the muscle action was explained in terms of Newton's laws. On the other hand, bodily functions such as digestion and metabolism as well as the application of medicine were treated as chemical processes. In accordance with the mechanistic paradigm, classical biology tried to reduce all aspects of living bodies to the physical and chemical interactions of their smallest constituents.

The inherent weakness of classical biology became exposed soon since it failed to explain the phenomena of growth, self-maintenance, reproduction and adaptation of organisms. Moreover, the model of machine is not suitable for interpreting the purposive and creative behavior of various organs in the body. The proposal that life is an *epiphenomenon* of material structure has to be abandoned in view of the special aspects of life.

In this situation, it was a major discovery in nineteenth century that all animals and plants are composed of cells. Treating cell as the fundamental unit of life, biologists realized that the components and functions of cells determined the structure and development of visible tissues and organs of an organism. The organization of a cell can be compared to that of a factory; hence the description of the function of cell requires the input-output model. It involved a shift from the mechanistic view to the process view; the new phase of biology is aptly called modern biology.

The unique feature of **modern biology** is that it describes the functions of various components of cells, instead of addressing the general questions about the structure of bodily parts. The shift of focus from structure to function is achieved through a new framework called machine-algorithm model upholding the *physical process view*. To clarify this point we can note that the functional units of a cell are made of different molecules such as protein, glucose, fat, water and DNA; these are compounds made of carbon, hydrogen, oxygen, nitrogen and few other elements. These molecules have combined to form the parts of a cell; it is the machine part. The interrelated activities of cellular parts as a whole can be treated as an algorithm.

Hence, the method of modern biology basically consists of describing the activities pertaining to life happening in a typical cell without enquiring the origin and existence of life as an entity. This has a special advantage as it can circumvent the question whether life is a separate entity that can be contrasted to the material processes of the cell. The various levels of an organism – cells, tissues and organs – have mechanical structures which would function according to predetermined pattern or algorithm. Accordingly, the term *life* refers to the totality of activities happening at various levels of physical body of organism.

As per the foregoing elaboration of the different worldviews and theories about the phenomenon of life, there is no unanimity with regard to the question: what is life? Consequently, the definition of life is a puzzle which has not been solved in the hitherto history of science and philosophy. We shall endeavor to get a coherent solution to this puzzle in the course of this chapter. As the first step towards this objective, we now turn to the key aspects of modern biology and its controversial themes.

2.2 Modern Biology

The foundation of modern biology is the **cell theory** proposed in 1839 which says that all forms of organisms are composed of cells and each cell functions as the basic unit of life. Another principle of cell theory is that all cells come from pre-existing cells by division, i.e., a new cell cannot be formed spontaneously and originally from inanimate things. Then, it follows that the beginning of an organism is in the form of a single cell which will be subsequently multiplied by the repetition of cell division. The number of cells of the organism increase in the order 1, 2, 4, 8, 16 and so on. Certain questions are to be answered for explaining the origin and growth of an organism: How do the cells of an organism get specialized to form different tissues and organs? What is the mechanism for coordinating the functions of different parts of organism? How are the hereditary features of organism passed on from cell to cell during cell division? What are the laws of heredity governing successive generations of organism? The answers to these questions will provide the key to understand the secret of life. Guided by this objective, modern biology has been developed into three complimentary branches -- *Cell Biology, Molecular Biology* and *Genetics*.

The average diameter of a cell is 0.02 millimeter (one by fifty of a millimeter). The body of a higher organism consists of billions of such cells. There are 100 trillion (10^{14}, which is 1 followed by 14 zeroes) cells in human body. There are wide differences among cells in terms of shape and size when we consider different organisms. Similarly, the cells in different parts of an organism also differ in physical features. There are exceptionally large cells also. A bird's egg is a single cell. Interestingly the egg of ostrich is the biggest animal cell and it has a diameter of about 18 centimeters.

The biological activities are performed at separate levels of complexity such as cell, tissue, organ and organ system. However, cell is definitely the basic level attracting our attention for discussing the puzzles of life. The different kinds of cells, with average size of 0.02 millimeter only, have a common structure formed by certain *functional units*. The important functional units of a cell are given below.

1. **Cell membrane**: It encloses the cell, defines its boundaries with other cells.
2. **Ribosome**: These are factories for synthesizing protein.
3. **Endoplasmic reticulum**: These units cooperate with ribosome for producing certain carbon compounds.
4. **Golgi apparatus**: It is responsible for storing and distributing protein and other products.
5. **Mitochondria**: These are the centers for producing energy through chemical reactions.
6. **Nucleus**: It is a structure existing at the centre of a cell and it is constituted by fibers called chromatin. The components of chromatin are Deoxyribo Nucleic Acid (DNA), Ribo Nucleic Acid (RNA) and protein which are macromolecules. When the cell is about to divide, the DNA molecules get condensed to become **chromosomes**. Thus a chromosome is entirely one DNA molecule.

There is a specific number of chromosomes for a particular species of organism. The number of chromosomes for some species is given below: human being - 46, rat - 42, cow - 60, dog - 78, cat - 38, mosquito - 6, cockroach - 23, rice - 24, sugarcane - 80. We will explain in the next section that chromosomes contain the genes carrying the genetic information.

It may be reiterated that the various functions of organs – metabolism, growth, response, reproduction and self-maintenance and so on – are ultimately performed by the functional units within the cells. *Cell specialization* is the key for the specific functions of tissues and organs. For example, the cells of intestines produce the hormones for digestion while the cells of liver produce bile and the cells of skin produces sweat. The particular structures of brain, bone and hair have separate characteristics because their constituent cells specialize in separate functions.

The diverse functions of cells are controlled by the macromolecules present in the nucleus, especially the DNA. In 1953, James Watson and Francis Crick discovered that every DNA has the form of a double helix, i.e., a spiral ladder. *It is hailed as the most important discovery in biological sciences.* Without going into the details of the molecular structure of the double helix model, we can say that DNA has two parallel sequences made of four kinds of nitrogen bases: adenine, thymine, guanine and cytosine (denoted by A, T, G and C respectively). In the case of human being there are 65 million bases (letters) in one base sequence of DNA. Since there are 46 chromosomes in a human cell the total number of bases for all chromosomes is 6 billion (65 million x 46 x 2 = 6 billion). In contrast, a bacterium has only 4 million bases in total. It is estimated that the total length of base sequences in all chromosomes of human cell is about two meters. It exists as coils in a highly condensed form within the nucleus, which typically has a diameter of 0.01 millimeter.

The central dogma of molecular biology holds that the biological activities of an organism are caused by the production of different kinds of proteins. DNA molecules contain the information for timely production of diverse kinds of proteins. This principle is expressed in terms of *gene* and *genetic code*, originally developed by M.W. Nirenberg, H.G. Khorana and other colleagues in 1966. Accordingly **gene** is defined as the sequence of DNA bases which corresponds to the production of a particular protein. In the nucleus of a human cell there are about 35,000 distinct genes which can produce about 50,000 different proteins required for the entire range of biological functions. The complexity of protein molecule can be inferred by noting that the molecular weight of a typical protein is 40,000 while that of water (H_2O) is 18. But there are proteins with molecular weights as high as 4 million.

It may be mentioned that the information for the production of a typical protein consists of about 900 bases or letters of A, T, G and C, including repetitions; this sequence of letters can be treated as a sentence. Now consider the sentence for the protein called *hemoglobin A*. The first fifteen letters of this protein sequence is GTGCATCTGACACCA. In a simple way, we can say that a gene is a piece of DNA which contains the information for producing a particular protein. So a gene has two levels. First is the physical level formed by the particular sequence of DNA bases; typically, there are about 900 bases in this level. Second level is the information (sentence) embedded in the particular segment of DNA. We are prone to think that such information accounts for a scheme, plan or purpose which is non-physical.

Now it can be inferred that the genes in the DNAs determine all biological activities of an organism through the production of proteins in various parts of body, in necessary quantities and at appropriate times. In a simple way we say that the genes carry the hereditary factors which are responsible for the characteristics of an organism including all biological activities. This is the basis of the general notion of a gene as the unit of heredity. The term **genetic code** is used for referring to the totality of genes included in the entire set of chromosomes found in an organic cell. Certain genes are found in two or more chromosomes; such repetitions of genes are also included in the genetic code.

As mentioned earlier, human cell contains 35,000 kinds of genes. Taking into account the repetitions of genes we can estimate the number of bases required as 180 million. This works out to be 3% of the total 6 billion bases. The remaining 97% of DNA bases do not carry the information for protein production; so they are treated as *Junk DNA* implying that they are the sequence of meaningless words. The totality of base sequences containing both active genes and Junk DNA pertaining to all chromosomes is called **Genome**; it is the interesting subject of study in recent decades.

For sequencing human genome, a multinational program called Human Genome Project was started in 1990. It achieved the complete sequencing of entire genome (6 billion bases in 46 chromosomes) by 2005. Main findings of the Project are given below.[# 5].

- The number of genes (about 35,000) in a human cell is only slightly greater than the number of genes in lowly organisms like roundworm and fruit flies.
- Human genome is 98% identical to that of chimpanzees.
- The genes responsible for particular diseases could be located contributing greatly to medical research.

The biological functions of all organisms living upon earth are similar; it consists mainly of metabolism, growth, interaction with environment, reproduction and self- maintenance. But biological world can be divided into numerous different species, where each species consists of organisms with certain similarities. So there are marked differences between the physical features of individuals belonging to two different species. Thus we can contrast between cat and dog. Additionally, there are conspicuous differences between two organisms belonging to the same species. In this way, we can observe the dual aspects of similarity-variation between species as well as within a species. **Genetics** is the branch of modern biology meant for explaining such aspects and its inheritance from parents to offspring. At the same time, it purports to give the fundamental principles behind the various biological functions of an organism, which may be grouped as under:

- Reproduction and Heredity
- The activities of cells
- Genetic code and protein production

The growth of a biological body is due to the *production of proteins* and it is manifested in the growth of bones and muscles.

Besides, it can be shown now that whole biological activities within an organism are performed by various kinds of proteins. Enzymes are one type of proteins; they accelerate and control the chemical reactions in the body. Special kinds of enzymes cause cell division. Another type of proteins becomes the building materials for prominent structures of body. Hair, skin, feathers and bones are made of specific kinds of proteins. Some kinds of proteins form filaments that cause the expansion and contraction in muscles and other movable structures. An important class of proteins is called hormones, which carry signals between different kinds of cells in the body. They serve as the link between consecutive nerve cells and thus cause the function of nerve systems as well as produce mental states.

In this manner, genetics reduces the functions of various organs of body into the production of proteins in the cells. As mentioned earlier, there are about 50,000 kinds of proteins in human body. The main task of *molecular biology* and *genetics* is to explain how the production process of proteins in the billions of cells is caused and regulated in a fundamental way. In this connection the structure and function of RNA molecules situated in the nucleus of a cell also needs to be mentioned. RNA molecule is like one strand of DNA double helix with the only difference that Thymine (T) is replaced by Uracil (U). This single base sequence functions as a messenger in the process of protein production. When a particular protein is to be produced, its code (a sentence consisting of about 900 bases) will be copied on the RNA. This messenger RNA goes to the outside of the nucleus and passes on the information to the ribosome where the concerned protein will be produced. Another point to be noted here is that the mitochondria - situated outside the nucleus and meant for producing energy - also contain some DNA that makes about 1% of the total cellular DNA. Such extra-nuclear genes produce certain proteins.

The above materialist process of protein production is summarized in the ***central dogma of molecular biology,*** which is expressed by the following two sets of statements:

1. *DNA makes RNA and RNA makes protein. The flow of information from DNA to RNA is called transcription and that from RNA to protein is called translation.*
2. *The DNA is able to replicate itself.*

This central dogma in the line of physical process worldview eschews our genuine question: what is the agency for controlling protein production and related biological activities?

The meaning of central dogma is that all biological activities are fundamentally caused by the DNA and the *information* embedded on it. Hence, *a gene has dual parts* – the material part consisting of a segment of DNA bases and the non-physical part called information for producing a particular protein. In spite of this fact, the idea that gene (or genetic code) denotes the information contained in the DNA has become popular in ordinary conversation and scientific literature; we must recognize that it is only a way of language.

It is clear that modern biology and its central dogma underscore the **machine-algorithm model** for studying the functions of an organism.[# 6] [*] There are many levels to this model. First is the level of genes, where the DNA segment is like the machine and the information to make a protein becomes the algorithm. As the second level, the entire DNA bases and information can be considered. Thirdly, at the level of cell, the material molecules of various functional units constitute the machine part, while the sequence of interrelated activities forms the algorithm. This model can be raised to the levels of tissues, organs and entire organism in a hierarchical manner. We may add that the machine-algorithm model or computer model is a mechanical and physical way of describing the aspects of life. In the context of this model, the term *information* is the physical version of genetic code.

Next step is to take up the philosophical enquiry -- epistemology and ontology -- about the concerned propositions of modern biology.

2.3 Criticism on Computer Model of Life [# 7][*]

This is in continuation of the philosophy of quantum physics presented in the previous book *Origin of Universe*. With regard to modern biology, firstly we consider the epistemology or theory of knowledge consisting of four parts -- methodology, source, justification and truth. After wards, the aspects of ontology will be presented. *The moot point is whether the scientific propositions of modern biology can be analyzed following the path of quantum physics.* For avoiding repetition of ideas already elaborated, only the specific issues of knowledge about life will be examined here.

Methodology and source

The laws about cause-effect relations and other sensible properties of cells as well as its components are to be analyzed in this stage. It is helpful to recall a few key points to familiarize with our new approach. The scientific laws are generally obtained through the sequence of theory (Ty), hypothesis (H), deduction (D), testing (T), and inductive inference (I). These stages are ordered like the organs of an animal. The propositions coming under successive stages of Ty, H and D together is called *deductive propositions (DP)*; the propositions of T and I are collectively designated as *inductive propositions (IP)*. We have suggested the new phrase *TyHDTI scheme* to denote the scientific method of combining DP and IP. In this context we can list the connotations of the four components of epistemology as following:

- *Methodology* is the deliberation about the general components – *Ty, H, D, T and I* -- of scientific method as well as about the relative importance of DP and IP in the meaning of scientific laws. Also, the various definitions and meanings given to the fundamental terms are to be clarified in different theoretical situations.

- *Source* denotes the theory about the structure of scientific mind, which generates the diverse kind of propositions under DP and IP.

From the previous section, we can state that the theory part of modern biology consists mainly of the following: definitions of cell and its functional units, cell theory, gene, genetic code (information), central dogma and fundamental ideas of genetics.

Scientists cannot observe life directly; it is inferred indirectly from the visible manifestation of biological activities such as growth, metabolism and reproduction. This implies that genes and genetic code are *unobservable* theoretical entities. At the same time the conception of DNA bases is rooted in deeper theory of quantum mechanics pertaining to subatomic phenomena, which are also unobservable. It may be reiterated that the *theory* of modern biology consists of the principles of quantum mechanics together with the concept of genetic code. In this situation, we can recall that the methodology of quantum mechanics is *logical positivism.* **So we must enquire whether the methodology (TyHDTI scheme) of modern biology is governed by the tenets of verification principle advocated by logical positivism.**

The definition of various genes is verified through experimental evidence and the resulting inductive laws serve to explain biological phenomena. Biologists regard the methodology involving the translation of invisible genes into biological activities, as highly successful. They cite scientific advancements like biotechnology, genetic engineering, gene therapy and cloning. Such wonderful achievements are projected as proof for the suitability of verification principle and logical positivism in biological research.

But it is to be admitted that biological phenomena exist as a higher level above the physical phenomena pertaining to subatomic particles, basic forces, atoms and molecules. Where as physical phenomena show the matter-energy duality, biological phenomena involve the duality of macromolecule (DNA, protein, etc) and genetic code. That is, biological activities show the interplay of physical and nonphysical aspects. In this situation, we have to adopt the idea of *layered universe*, which necessitates the conception of different levels of theories. Though modern biology adopts the physical process view whereby the nonphysical aspect of genetic code is reduced to physical behaviors, this approach has serious defects. **The following points *will explain why the computer model of life cannot be accepted*** as the paradigm for studying biological phenomena.

One. A corollary of the central dogma of molecular biology is the principle: *one gene- one protein*. It means that the information contained in a particular gene is uniquely responsible for the production of a particular protein. But on the basis of the Human Genome Project and the methods of Bioinformatics, it is now clear that the relation between single gene and single protein is not precise. There are situations where many genes cooperate for producing a particular protein. More over the location of a gene is not fixed in the sequence of DNA bases. One alternative form of a gene is called *allele*. The manifestation of multiple alleles has been detected and it adds to the confusion about the identification of genes. Another perplexing situation is that a given gene is split into many pieces which are situated in different parts of the chromosomes. We can see a sort of joint action involving many genes for the production of related proteins. These aspects deny the possibility of translating an invisible gene to observational terms, implying the failure of verification principle.

Two. It is well known that there are many feedback mechanisms within a cell so that the gene theory is not sacrosanct. The existence of mitochondrial DNA outside the nucleus of a cell is not explained satisfactorily. Similarly, the opposite forces of gene expression and gene regulation are well beyond the purview of mechanical algorithm. Most importantly the role of junk DNA that constitute up to 97% of genome is a matter of wild speculation.

Three. As mentioned above, biologists fail to determine uniquely the genes causing the particular functions of cells and higher organs. In other words, the phenomenon of life exhibit indeterminism and freedom in considerable extent. We have to admit that genetic engineering – biotechnology, cloning and so on -- is not purely mechanical since there are elements of nonphysical attributes also. This point explains why the concerned experiments have not succeeded contrary to the expectations of scientists. Additionally, it is pertinent to remember that the physical method of logical positivism suffers from the dangers of underdetemination because the vested interests of scientists would corrupt the theory so that the experimental method becomes biased.

Four. Modern biologists adopt the latest theory about mind named as *computer model functionalism* (CMF) which is in accordance with the principles of cognitive psychology. But the consciousness and other nonphysical aspects of our mind cannot be captured in the computer model – this point will be explained later.

Can We Justify the Propositions of Modern Biology?

Here *justification* deals with the issue whether the scientific law represents actually existing aspects of biological world. So the essence of justification is that the components of theory (Ty) as a whole represent the biological phenomena; we must get sufficient evidences for that judgment. It finally leads to the question: do macromolecule (DNA, protein, etc) and genetic code exist?

To clarify this point, consider the example of a physical object called fan. We can express the activities of fan using an algorithm or flow-chart, depicting the sequence of activities of the components of fan. As explained in the previous book, this machine-algorithm model is in process view; its justification involves the conflict between instrumentalism and naïve realism. Then we have to adopt content view about the existence of matter -- that is, naïve realism -- for the justification of knowledge under classical science and quantum mechanics. Since the issue is not settled at that stage, we proceeded to the field of quantum cosmology, which is concerned with the existence of the fundamental components of matter such as quarks, strings and membranes. Then the relevant theory of justification is ***scientific realism***, which suffers from the controversies of pluralism and skepticism.

In the similar vein, we have to investigate whether DNA and genetic code (information) exist as per the machine-algorithm model.

In the materialist way of knowledge, the existence of DNA-information duality requires that DNA exists as a particular organization of matter and it is independent of information. It boils down to the vital question: do matter and information exist? On the basis of Immanuel Kant's philosophy about *existence*, we can accept that the theoretical entities of modern biology are predicates only. Accordingly the theoretical entities do not exist in the form of separate objects. Thus, scientific realism is to be rejected. However, we apply Aristotle's rules of thought to scientific knowledge that deals with the physical and mental phenomena. This is the reason why modern scientists treat machine and algorithm as separate entities constituting the computer model. Considering these ideas we can summarily state that the theories of quantum physics are not competent to justify the doctrine of DNA-information as well as the propositions of molecular biology and genetics.

Theory of Emergence and Self-organization

Modern biologists prefer to adopt the notion of *emergence* in order to describe the origin of life in tune with the computer model. It means that, in a structured organization, new properties emerge at higher levels of integration which could not have been predicted from knowledge about the lower-level components. For example, the wonderful properties of water emerge when the atoms of oxygen and hydrogen combine in a particular proportion. More familiar example is that of a company (firm), which is a higher level of organization above a set of entrepreneurs. The company has distinct properties as compared to the personalities of individual members.

Scientists like Fritjof Capra have tried to popularize the idea of emergence of life by linking to the aspect of *self-organization*. They argue that life is an emergent property of the particular organization of cell. When certain macro molecules (material compounds) like DNA and proteins are organized in a peculiar way, the property of life emerges. Then the question arises as to what is the cause of organization of the first cell. Here, adhering to materialist path, the proponents use the concept of *self-organization,* implying that the macro molecules are not guided by any metaphysical force. [# 8]

It can be criticized that, in the context of organism, the above concepts of emergence and self-organization are proposed from empirical perspective – it is mere description of the sensible properties of organism adopting physical process view. This can be challenged through the problem of justification. Can we say that a physical organization, like company or cell, exists actually? This question leads to skepticism as famously articulated by David Hume.

To say it bluntly, the theory of emergence and self-organization together is another version of machine-algorithm model. Here the proponents would adopt the descriptive view about the hierarchy of living beings to say that life emerged from inanimate macromolecules like DNA. We have just noted that the existence life cannot be conceived in this process view; so origin of life is not known. The foregoing epistemological analysis of modern biology throws up serious dilemma in the ontological deliberation. Now we may proceed to solve this problem in an innovative manner.

2.4 System Philosophy of Life

It is expedient to reiterate the main drawbacks of scientific explanation about the origin of life. *Naturalism* tries to treat life as an epiphenomenon of material evolution. This is the paradigm of classical biology and related sciences. Earth was formed about 4.5 billion years ago when it consisted of low level atoms only. During the first hundred years, the inanimate atoms called hydrogen, oxygen, nitrogen and carbon combined in various stages to form macromolecules like DNA, RNA and protein. These macromolecules somehow got the power to organize itself for constructing the structure of a cell. So, according to scientists, life emerged in the first cell about 3.5 billion years ago as a byproduct in the process of inanimate evolution. However, there are many philosophical drawbacks in the description of the evolution of earth as well as the doctrine of *epiphenomenalism.* Scientists are at wits end in explaining how life with such wonderful non-physical properties can arise from inanimate macromolecules.

When modern biologists hold that information (genetic code) is an algorithm, they treat it as a set of mechanical rules similar to the software of a computer. Then the challenge facing us is to explain how DNA and genetic code exist. In the case of a familiar object like fan or computer, the algorithm is the product of the mind of manufacturer or programmer. Considering the philosophical problems associated vitalism and spiritual versions, which are metaphysical theories of life, we do not import the notion of God or soul in this treatise. For explaining why genetic code is radically different from a computer program, it is possible to show that the information contained in DNA is not purely mechanical. In this context, we observe as following:

- ❖ The functional units of cell display certain purpose and creativity at the fundamental level. Different kinds of tissues and organs are formed by the specialization of cells. Scientists cannot give satisfactory answer to the question: how does cell differentiation take place when the genetic code is the same in all cells? Since thousands of genes are engaged in protein production at the appropriate times and in suitable measures, they display the dual factors of competition and cooperation – it obviously involves self-organization. Taking into account these evidences, it must be admitted that the genes have a sort of *consciousness* about time and purpose. Physical body of an organism is a Great Miracle, much beyond the scientific realm of human understanding.
- ❖ In the case of a cell, the factors like creativity, freedom and opposite purposes must be treated as non-physical. We cannot see such properties, when we assume under physical view that cell is an assembly of material components. *So we are inclined to hold that the genetic code is not physical or mechanical; it is in sharp contrast with the software of a computer.* Here we follow the path of scientific thinking without the influence of religious concepts of God and soul. It may be added that modern biology under physical process view gives only the description of various

activities happening in biological organism. Obviously, it is the linguistic approach to scientific knowledge, ignoring the traditional question of existence of theoretical entities.[# 9].

*An important implication of the philosophical problems of physical process view is that we must shift to **content view** of knowledge for conceiving the theoretical aspects of biological phenomena.* Taking into account the above critical points, we will now introduce a new philosophical scheme named as *System Philosophy* primarily to explain the existence of life in various cells and higher organisms. The central concepts introduced here are *system, phenomenal existence* and *real existence* which pave the way for a new understanding of biological organism.

2.4.1 Philosophical Definition of System

At present, we are concerned with the fundamental components of a biological cell, namely DNA and genetic code, which display the physical and nonphysical properties. When these theoretical entities are considered as separate entities, they are mere predicates without existence. Then the only alternative is to show that DNA and genetic code have phenomenal existence in an interconnected manner. As modern biology is a phenomenal knowledge, its theory of justification also must be phenomenal; we want to discard all kinds of realism. We can tide over these problems by postulating that the theoretical entities are interconnected since they represent systems. We philosophically define **system** as a productive structure composed of opposite entities. System has existence due to the axiomatic properties presented below.

*a) **System** is a **symmetrical structure of X-Y coordinates**.*

The opposite components of a system are similar to the X axis and Y axis pertaining to coordinate model of analytical geometry. Here X and Y have dialectical and productive relation – it is the essential aspect of system.

b) Production Function model.

System has a production function similar to that of a factory. According to economics, factory is a system for producing various goods using the inputs called labour and capital. The production function model of factory is well known where X axis is labour and Y axis is capital. The production of various outputs with different combinations of X and Y as inputs is illustrated on the basis of analytic geometry. [# 10][*].

c) Complementarity of opposites.

As a rejection of Aristotle's rules of thought, the opposite entities X and Y are not independent entities. Instead, X and Y are opposites, which have complementary character. We cannot define one entity without considering its opposite entity. The opposite components of a system are complementary to each other; they have interdependent existence. Accordingly, **it is postulated that a system is a whole of opposites and that only system has existence.** We can talk about the opposite components, then they are predicates only – such opposites do not exist separately. Since system or factory has existence, its output also can be said to have existence.

In the X-Y model there are four blocks, technically called as *quadrants*. We can recognize that things have the opposite qualities of good and bad. While, the good things occur in first quadrant denoted by (+X, +Y), the bad things appear in the third quadrant denoted by (-X, -Y). It may be clarified that a particular thing has both states of good and bad depending on the circumstances. Snake's venom is good in some situation while it is bad in other situations. This consideration of good and bad pertains to our knowledge of value; hence, the system model of four quadrants has special relevance in the subject of ethics. However, the outputs of a system are factually marked in the first quadrant (+X, +Y), without the distinction of good and bad.[# 11].

2.4.2 Cell is a Phenomenal System

We can logically apply the principle of system to describe the phenomenal existence of organic cell and its components. At the outset the structure of an organism is divided into visible level and invisible level. Then there are two sublevels for visible level. These facts are shown in the table given below.

Table 2 : Levels in the Structure of Organism [#12][*]

Visible level	3. The level of cells (cells, tissues and organs) 2. Functional units in the cell (membrane, ribosome, nucleus, chromosome, DNA, RNA, protein etc).
Invisible level	1. Genes (segments of DNA and information).

DNA is the main macromolecule existing in a cell containing information or genetic code which is a nonphysical aspect. Similar opposite aspects may be found in other macromolecules like RNA and protein also. *Considering this fact, it is better to take macromolecule as a generic term for representing the physical aspect of life.* We cannot separately observe the macromolecules and information. This is the reason why they are included in the invisible level of the above table. Next stage is to hold that macromolecule and information respectively represent the opposite forces of *matter* and *consciousness*. For the time being we call the opposite terms as macromolecule and information only, because they refer specifically to the subject of modern biology.

The most important issue to be settled in this context is about the source of the nonphysical aspect called genetic code (information).

The empirical scientists who brood over this question are sharply divided into two groups. First is the group of materialists adhering to the physical interpretation of life; they do not like to see the nonphysical aspects of organic cell. The second group resorts to the notion of intelligent designer existing outside natural world. We can say that they try to combine the metaphysical vitalism with the empirical naturalism; this mixing of two worldviews is bound to create intractable problems of thought. A detailed discussion of *Intelligent Designer Argument* is postponed to the next chapter because the concerned issue becomes more explicit in the context of biological evolution. Here it suffices to say that the controversy about intelligent designer can be overcome now by holding that the nonphysical aspect is inherent in the organic cell itself. The notion of system accomplishes just that.

Chapter 2/Diagram
System Model of Organism

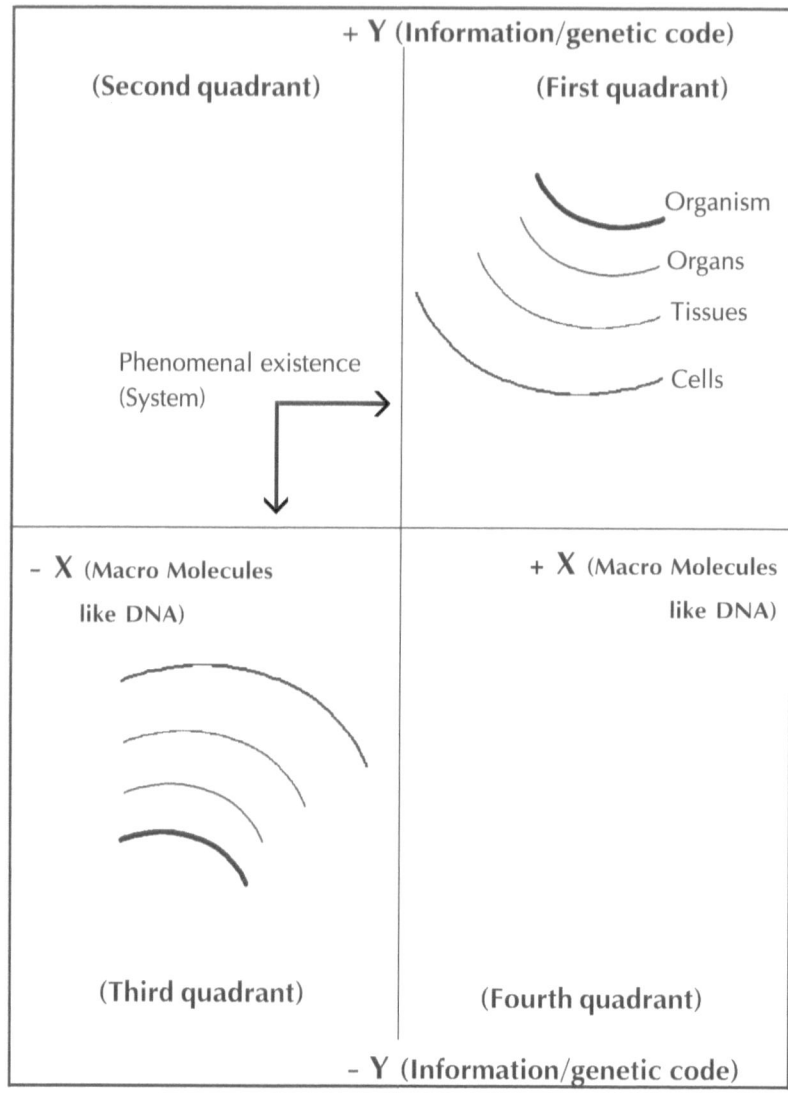

The opposites called macromolecule and information can be arranged in the form of X and Y coordinates. This system has the function of producing the invisible and visible levels of organism. In our ordinary way of speaking we can say that DNA, RNA, proteins and the various functional units of cell are different levels of macromolecule-information duality. The X-Y coordinate system of macromolecule and information has *phenomenal existence*. At the higher levels, we can see the systems called cells, tissues and organs which hierarchically constitute an organism. The diagram showing the structure of an organism in this manner can be called **System Model of Organism**.[*].

Here it is necessary to clarify the distinction between the *system model* and the *computer model*. The system model is under content view, which holds that DNA and other macromolecules exist as wholes with physical and nonphysical parts. But we ordinarily treat physical part and nonphysical part as separate objects, as a consequence of Aristotle's rules of thought.

The empirical approach of classical biology is to reduce the non-physical aspect into physical properties so as to uphold epiphenomenalism, in accordance with naturalism. As a further step, the method of physical process view adopted for modern biology – the computer model – treats the nonphysical aspects of organism as a physical algorithm. It may be reiterated that the purpose of science is the production of practical knowledge about various phenomena, by reducing the non-physical aspects of system into physical properties. In this context, we may recall the earlier criticism on the empirical notions of *emergence* and *self-organization* also.

In the computer model of life, macromolecule and information are supposed to exist as separate entities. This generates the problem of skepticism. However, the principle of phenomenal existence of system, introduced above, solves this serious problem. Cells and other levels of organism have existence in the phenomenal sense because they are formed by the systems of macromolecule and information. This principle serves as the *justification* for the propositions of biological sciences.

What is Life?

Now we can present the systematic answer to this ontological question in the following steps:

- Life is not a metaphysical being or substance that is separate from physical body.
- Cell is a peculiar organization of material compounds, technically called macromolecules. This cell has life. We should not say that macromolecules organized themselves to form the first cell and then the property of life emerged. Instead, we have to define life as the aspect of organization of cell itself.
- *Life exists as the system* of opposites namely physical macromolecules and nonphysical information (genetic code). In other words, life is a higher level of organization of specific macromolecules, mainly DNA and protein. To explain this fact, *we can say that life is like a company or firm constituted by a group of entrepreneurs.* It is a pattern of relations that has existence as per the system model under content view.

Now let us describe life adopting process view. Modern biologists popularized the emergence theory that life originated in the macromolecules like DNA; but this physical process view is refuted now. Through the dialectical process of macromolecules and information (X and Y co-ordinates), the first cell was formed as a system. The properties or activities of this cell – such as metabolism, growth, reproduction etc – are denoted by the word 'life'. *We should not construe life as a separate entity different from the physical structure of organism.* [*].

It is enlightening to note that the system model of organism can be linked to the existence of inanimate world. We will show later that by extending the application of the principle of *system* to material things we can present the unifying vision of all levels of natural phenomena. Thus we realize that System Philosophy provides the framework for studying the structure of our universe.

The hierarchy of organisms in the biological world is illustrated comprehensively using the system model of organisms. For that purpose, we have to distinguish between various systems. A suitable modification of Aristotle's rules of thought is proposed here: the items A and B mentioned in the three laws are systems; then A and B are not predicates. The depiction of hierarchy among organisms such as bacteria, plants, fishes, reptiles, animals and humans is called *The System Model of Biological World*. The hierarchical levels of organisms are now illustrated as the products of increasing complexity on account of the inputs called macromolecule and information. We will utilize this seminal point to address the philosophical issues arising in the theory of biological evolution to be presented in the next chapter. [*].

NOTES of Chapter 2

1   **The main references** used for this chapter are: Alan Grafen and Mark Ridley (2007), Ammar Al Chalabi, et al (2007), Behe (1996), Bird (2003), Brennan (2005), Capra (1983), Capra (1997), Chardin (1965), Darwin (1859), Dawkins (1976), Dawkins (2007), Dawkins (2009), Dennett (1991), Griffin (2000), Guttman (2007), Guttman et al (2006), Haught (2000),Haught (2001), Heil (2003), James et al (1987), Jantsch (1989), Job Kozhamthadam (editor) (2004), Kant (2003), Lewens (2007), Leahey (2005), Mayr (1999), Miller (Editor) (2001), O'Leary (2004), Robert John Russell (Editor) (2004), Shaffer (1994), Tarnas (1991), Thomas (General Editor) (2012), Vijayakumaran Nair and Jayaprakash (2007).

2   The Table of worldviews is an original idea presented in this book and it is the basis of our philosophical analysis.

3   The distinction between rational and empirical approaches to the process view of life is highlighted here. It may be noted that evolutionary theology is a rational theory of evolution. See the concerned articles of *OMEGA – Indian Journal of Science and Religion* (Institute of Science and Religion, Aluva, Kerala).

4   The brief discussion of spiritual science, paranormal phenomena, parapsychology and pseudoscience given in previous book *Origin of Universe* is recalled here.

5   Vijayakumaran Nair and Jayaprakash (2007), *Cell Biology Genetics Molecular Biology, page 356.*

6   The linking of central dogma with the *machine-algorithm model* is my original idea.

7   The epistemological criticism of computer model of life is an original presentation; it is articulated on the basis of the philosophy of science explained in various parts of earlier book.

8   Empirical description of emergence and self-organization is given in Mayr (1999), Capra (1992), Capra (1997) and Jantsch (1989).

9   Treating genetic code (information) as non-physical is the vital step to reach the System Philosophy of Life, to be introduced in the next section.

10   It may be mentioned for record that my philosophical idea of *system* germinated from the study of economics, mainly production function of firm, which I undertook in 1990s.

11   I am indebted to the books of Fritjof Capra in the deliberation about *whole* and *complementarity of opposites*. Capra described *system* in an empirical manner to show the features of interconnectedness of opposites. But the representation of a philosophical system using the X-Y coordinate model is my original idea. This marks the birth of System Philosophy.

12   The layered view of organism is the crucial step to develop the system model of life.

Chapter 3

System Philosophy of Evolution

****

3.1 Organic and Mechanistic Worldviews

3.2 Spiritual Process View of Evolution

3.3 Physical Process View : Darwinism

3.4 Fallacy of Intelligent Designer Argument

3.5 The System Model of Biological Evolution

Author's main original ideas are marked by [*]

The mark [#] gives the number of note at the end.

****

The most hotly debated issue in recent centuries is about explaining the wonderful diversity and complexity of living beings. The scientific and religious views on this matter are diametrically opposite; it is the main reason for the controversy. In this situation we intend to explain the relevant aspects of the scientific theory of evolution so as to make an epistemological analysis. Subsequently, it will be explained that the opposite camps of evolutionary debate are engaged in an emotional and ideological battle from slippery positions. Happily, with reference to the System Philosophy of Life introduced in previous section, we can anticipate a new philosophical doctrine of evolution for synthesizing the different worldviews.

We need to distinguish between the words *change* and *evolution* appearing in our ordinary usage. Everything in our nature has got permanence and change – these aspects are opposites which we experience daily. When we see an object firstly we become aware of its permanence. But in the course of time, the object undergoes change. A plant grows and it withers; this is an example for change. But evolution is a specific type of change. Evolution can be defined as a change to a higher level of complexity. Development and purpose are embodied in it in a special way.

In the context of organisms on earth, the topic of **biological evolution** is an important field of enquiry in science. We try to understand evolution scientifically using the framework of biology, which holds that life exists basically in cells. Scientists have discovered that life originated in the first cell (bacterium) about 3.5 billion years ago. Later the bacterium multiplied and the process of evolution started. Thus plants, fish, amphibians, trees, animals and human beings were formed as the different levels of biological world. Each level can be divided into further sublevels. In this classification scheme, *species* is the most important category; it is defined as the group of organisms which can interbreed, but are reproductively isolated from other such groups. Horse, dog, cat and monkey are examples of different kinds of species in the animal world. Today more than two million species inhabit upon earth. The process whereby such a great number of species, starting from the stage of single cell, were formed in the course of about 3.5 billion years is called biological evolution.

In the early stages of evolution, living beings had less number of cells; the number of species were also few. Later living beings as well as species increased in number and achieved more complex features. Thus in the flow of time higher levels of species evolved from lower levels. It means that certain species underwent changes so that more complex species emerged. We may recall the Table of **Worldviews about Life**, given in earlier section 2.1, in order to deal with the topic of biological evolution.

3.1 Organic and Mechanistic Worldviews

The organic worldview (idealism or **vitalism**), which became a religious philosophy called theism, envisaged the idea of *fixity of species*. Accordingly, all species that we see today existed from the very beginning. This belief was adopted by the major religions like Judaism, Christianity and Islam. God created the levels of complexity in the species. *There is no notion of biological evolution in theism.* The story of creation included in Bible upholds the notion of fixity of species.

During the period from 16th to 19th centuries, the mechanistic worldview (empirical) and classical biology treated organisms as machines made of material atoms. In this situation, the ontological theory of classical biology is **naturalism** and it is an extension of materialism. This theory was concerned with the static existence of different species and organisms, without involving the notion of evolution. On the contrary, the spiritual process and physical process worldviews recognized the aspect of biological evolution – these doctrines are briefly given in the following sections.

3.2 Spiritual Process View of Evolution

The romanticism flourished in sixteenth century and afterwards. It asserted that the existence of a vital energy in living being is the manifestation of the immanent God. The divine energy caused the evolution of physical bodies leading to the formation of complex species. In the recent history of this metaphysical process view, Teilhard Chardin (1881- 1955) and A. N. Whitehead (1861-1947) presented mystical and theological insights about biological evolution; such doctrines can be collectively named as ***evolutionary theology*** or *process theology*.

Here, we have to specifically consider the efforts made in the period between 1750 and 1850 to link the *mystical process worldview* with the science of biology. The rapid progress of industrial revolution and the construction of new rail links led to the discovery of fossils. As a definition, fossils are the remains of prehistoric plants and animals deposited in the layers of hills and rocks. The fossils unearthed while cutting the hills sides to pave the way for new rail lines became the key assets of research in natural biology.

The study of fossils by George Buffon, Erasmus Darwin and Lamarck (1744-1829) provided great proofs about biological evolution. In this context, the theory of romantic evolution proposed by Lamarck in 1809 is the most important development. It was an effort to link the mystical process view with the evidences of fossil-science. Later this theory came to be known as **Lamarckism**. In adverse circumstances the immanent vital force caused certain changes in the body of organism, so that it could adjust with the environment. These bodily changes are termed as *acquired characteristics*. When such acquired characteristics are passed on to next generations they become *hereditary* features suitable for adaptation. By the accumulation of these adaptive features over many generations, a new species is formed.

The working of vital energy as described above typically involves metaphysics. The example of giraffe is commonly used to explain the romantic theory of evolution. The ancestors of giraffe did not have long necks. To eat the leaves which were very high on the trees, the ancestral animals used to stretch their necks – then long neck became an acquired characteristic. When this effort continued through generations, the neck of offspring got elongated due to the influence of immanent vital energy. In this way, the acquired characteristic became hereditary; it explains the origin of a new species called giraffe.

The mystical theory of biological evolution as outlined above has serious drawbacks. *Firstly*, the idea of immanent vital force involves realism about metaphysical forces; but we have already presented philosophical arguments for rejecting realism. Moreover, science has no proof to establish the existence of supernatural powers and their activities; these are beyond the scope of science. In short, there are many philosophical problems in describing how metaphysical beings work on physical body. *Secondly*, Lamarck's hypothesis that acquired characteristics are transferred to the next generations cannot be accepted as true. Can we generally say that the musical, acting or driving skill of parent will be inherited by son or daughter? It may be remarked that the scientific laws of heredity were unknown in Lamarck's time.

3.3 Physical Process View : Darwinism

In the second half of 19th century, the study of biological evolution was shifted to the realm of **physical process view**. According to this new paradigm, science can study the changes happening in the bodies of some members of a species by observing specific physical causes. This approach of scientific study begins from **Charles Darwin** (1809-1882) who created a whirlwind in modern thought. The ensuing paragraphs will discuss the key aspects of Darwin's theory and also present suitable criticisms about his revolutionary approach.

The path breaking book of Charles Darwin published in 1859 was titled *On the Origin of Species by Means of Natural Selection or the Preservation of Favoured Races in the Struggle for Life*. Though another naturalist called Alfred Russel Wallace (1823-1913) also presented a similar theory, Darwin's book gained wide publicity because it provided clear cut evidences using simple language for supporting the theory. In the *Origin of Species*, Darwin did not explicitly deal with the evolution of human species. Later he proposed that humans have descended from lower animals belonging to the class of apes and accordingly published his book *The Descent of Man* in 1871. The principles contained in these two books are collectively called Darwin's theory of evolution. [#1].

We can summarize Darwin's theory of evolution into four main parts – *variation, struggle for life, inheritance and natural selection*. Darwin treated these key terms as physical processes to be explained below. We may remember that the biological principles underlying his theory are known only in the 20th century.

1. **Variation** – Due to the influence of external factors there may be difference in the physical features among individuals in a species. Darwin treated such bodily variations as events of chance. He used the word *chance* just because he did not know how the external

conditions of life caused physical variations. (According to Genetics, a new branch of biology developed after 1920, the external factors cause mutations of genes existing in cells. Mutation is the sudden variation in the structure of genes and it may produce differences in the physical features among organisms).

2. **Struggle for life** – In living world, there is a tendency for proliferation of organisms within a species. At the same time, natural resources for the sustenance of life are limited. Under these circumstances, there is a struggle for life between individuals in a species. The principle phrased as *survival of the fittest* holds good here.

3. **Inheritance** – Some physical variations are helpful for the organism in its struggle for life; some other variations are not useful. Individuals in a certain species who get favorable variations in the body will live much longer. They will produce more off springs. It implies that those characteristics which help in the struggle for life will be passed on to next generations through inheritance. The individuals who do not inherit favorable characteristics will perish soon and they would not have successors. For example, consider a group of wolves. Those wolves having stronger legs will run faster and capture more preys; they can live longer and produce more off springs. The wolves that cannot run fast will not get preys and they will die with hunger. In this way, the property of strong legs are inherited through next generations. (Gregor Mendel discovered the basic laws of heredity in 1865. Subsequently in 20th century, genetics flourished with new discoveries through which the hereditary factors are called genes).

4. **Natural selection** – When there is a struggle for life, the individuals who get favorable physical variations will survive to produce more off springs. The individuals without favorable inheritance would die sooner. Herbert Spencer in 1852 expressed this principle as *survival of the fittest*. Darwin utilized Spencer's idea to establish that biological evolution occurs due to natural selection. We can simply define natural selection as the process whereby small favorable physical variations are preserved through future generations and consequently, in the course of time, new species is formed.

Darwin describes the natural selection, which occurs through several generations, in a materialist sense without involving metaphysical powers. Thus he was able to give a scientific foundation to the theory of evolution. Darwin's proposal that the various species are evolved from a common ancestor is his most far reaching contribution. He says that life developed on earth as per the model of a tree. The expression *Great Tree of Life* is a nice imagery. Later in 20th century, modern biologists have carved the stages of biological evolution by analyzing and grading fossils.

Now we will note that Darwin's theory is in the sphere of materialism. By physical variations, Darwin meant the changes occurring in the bodily structures of organisms. He felt that these physical variations are the result of just chemical reactions, without any purpose or creativity. According to science, there is no metaphysical soul in organism. It is to establish this point that Darwin imagined variations as *chance* events. **We can note that in order to explain the reasons for a sudden change scientifically, without involving supernatural forces,** *chance* **is the best word to use.**

Charles Darwin formulated his theory of evolution adopting the physical process view, which was based on the classical biology and naturalism as prevalent in the first half of 19th century. Hence, Darwin's postulates on physical variations were vague and weak. There are many critical points against Darwin's theory since it proposed that biological evolution occurred in a materialist way. The criticism can be condensed into three statements:

- If man was evolved from apes it is a great leap in evolutionary history. Darwin has not explained how it happened.
- Darwin was not clear about the links in the Tree of Evolution.
- Religious people consider that Darwin's materialist theory is a denial of God.

Later, modern biology was established with the development of cell biology, molecular biology and genetics. These subjects hold the key idea that cell is the fundamental unit of life; further, it aims to reduce the phenomenon of life into the activities of genes embedded in the various sequences of DNA bases. New concepts like DNA and genetic code are used to eliminate the vagueness of Charles Darwin's theory of evolution.

The refined theory of evolution is called **neo-Darwinism.** Now mutation is defined as the accidental change in any gene due to the pressure of natural circumstances or the activity of certain virus. When mutation occurs to certain genes, the physical features of the organism may undergo variation. When mutations happen to a group of genes, they display the tendency to sustain the favorable changes and to eliminate the unfavorable ones. As the division and proliferation of cells occur, more copies of favorable genes are taken and the unfavorable genes are discarded. This process can be described as the *competition of genes*. Accordingly the struggle for life takes place in the level of genetic code.

As per neo-Darwinism, inheritance is the flow of a community of genes through successive generations. Richard Dawkins, in his book *The Selfish Gene* (1976), gives special explanation about how good genetic mutations survive in the off springs. When the division and proliferation of cells occur, more copies of favorable genes are taken and they are expressed. This is qualified by the term *selfishness* of genes. It implies that there is selfish competition among skillful genes, which continues through inheritance. At the time of origin of embryo, the genes coming from father and mother get mixed up. During this initial stage, the genes which are capable of adapting with environment must succeed in the competition for survival; then only the favorable genes will become a hereditary factor. Then, *natural selection* is the process of forming a more complex species when the favorable hereditary characteristics get accumulated in the long course of time, through generations. Many external factors like geographical isolation, climate and variety of food greatly influence the formation of new species through natural selection. [# 2]

However, the modern postulates about biological evolution are still within the framework of Darwin. Therefore, Darwin's original theory and neo-Darwinism together is referred to as **Darwinism.** For deliberating upon the methodology of Darwinism or evolutionary science, we must consider the notion of levels of theories. Knowledge about evolution is firmly based on the findings of modern biology, mainly the genetic theory. Here, life is an algorithm (activity) of the macromolecules like DNA, RNA and proteins present in the cells of an organism. The algorithm exists in the form of genetic code. This is the computer model (machine-algorithm model) of life. The methodological doctrine supporting this model is *logical positivism*. Then Darwinism is a higher level theory as compared to the genetic theory of life.

Now we may mention a few important points to show that logical positivism and its verification principle are not properly applicable to the study of biological evolution. In the functions of genetic code as well as in the factors of evolution such as mutation, struggle for life, inheritance and natural selection we can observe many non-physical attributes. Certain interesting evidences accumulated from the contemporary study of evolution are highlighted below.

First. The evolution of higher order animals including humans have occurred during the last fifty million years and it is the timeframe for Darwin's theory. Actually, Darwin focused on the evolution of animals and certain varieties of birds because he had no clear evidences about the evolutionary history of preceding 3450 million years. Moreover, the species are clearly distinct without any apparent continuity of evolution. There are ongoing debates whether the animals are evolved from simpler beings about which fossil evidences are not satisfactory.

Second. It is in this context that the hypothesis of *punctuated equilibriums*, presented recently by Stephen Jay Gould, has gained prominence. Accordingly, in the evolutionary history there are long periods of equilibrium running into millions of years without any occurrence of new species. Then, suddenly evolution happens bringing out some new species; it is followed by equilibrium period again. So, as Jay Gould says, evolution with marvelous types of biological diversity occurred in the gaps or punctuation marks between successive long periods of equilibrium. The fossil evidences support this point of view. In that situation, the Darwinist theory of small variations does not hold true. Biological evolution is not a continuous process. The physical process view is incapable to explain the sudden and miraculous emergence of new species. For example, scientists cannot explain how bacteria evolved into plants and how fish evolved into reptiles and animals. We do not say that a car is evolved from motorcycle by adding two more wheels, joining small parts repeatedly and forming an outer body.

Third. The serious controversy about **missing links** may be considered now. The creationists argue that if a higher species was evolved from a lower one, through a series of small variations, then the fossils of the intermediate levels also must be available. But fossil records do not satisfy this condition. This is the *problem of missing links*. It will be explained later that there are some serious misconceptions in this problem and it can be removed by adopting the content view about biological evolution. The missing links actually denote the gap and jump occurring in the evolutionary sequence. It may be remarked here that the above mentioned theory of punctuated equilibriums also helps us to address the issue of missing links.

Fourth. The Cambrian Explosion, happened around 600-500 million years ago, is a wonderful episode. A group of evolutionists advocate that the original forms of all latest species simultaneously existed in this period. This proposition would pose a serious challenge to Darwinism and its assumption that the species evolved in hierarchical succession. Similarly, there is controversy related to the extinction of dinosaurs, which lived upon earth during 200-100 million years ago. The miraculous turn of events in the evolutionary history cannot be explained within the framework of materialist Darwinism.

Fifth. There is a remarkable mystery regarding the pace of mutation in the history. During the first 3 billion years after the emergence of life, the evolution was very slow. Subsequently, the speed of the process increased. The recent fifty million years accounts for most of the diversity of species. More importantly, human race attained the present form only about 40,000 years ago. The acceleration of evolutionary sequence is a secret that goes beyond the premise of Darwin's theory.

In addition to the above points about mutation, we can deliberate upon the other factors of evolution also. The *struggle for life* (survival of the fittest) essentially includes the purpose of maintaining life and taking it to higher ends. Besides, as we have mentioned earlier, the *inheritance* involves the selfishness of genes and such strategies reflect purpose. Moreover, *natural selection* is not a blind process of chance, but is the purposive accumulation of favorable features which have the potentiality to transform into a new species. Without going into the details, we can say that the last nail on the coffin of logical positivism is hammered by the idea that genetic code is not simply a physical algorithm since it has many non-physical attributes like purpose, creativity and freedom. As noted there, the existence of macromolecules and genetic code cannot be established due to the skepticism afflicting the justification theory called *scientific realism*. The inferences and laws of evolutionary science lack justification because there is no way to show the existence of macromolecules and genetic code as distinct entities.

However, evolutionary scientists claim that Darwinism is the only feasible theory about the origin of species, since this theory is adequately supported by evidences from fossils and geological data. The attitude of *scientism* and the self-congratulating assertions of biologists stem from a sort of language game that dubiously treats Darwinism as the only explanation of biological evolution. We need a wider and deeper perspective to see the nature of truth in scientific propositions about evolution. The factors of evolution as a whole are better expressed by the term ***creative evolution***; it opens up the gate for a higher philosophy using the concept of system. [# 3][*]

3.4 Fallacy of Intelligent Designer Argument

In recent decades a group of biologists have argued that the nonphysical aspects of organism have come from an Intelligent Designer, external to natural world. This idea is proposed on the basis of empirical knowledge about world and it is viewed as a compromise between naturalism and vitalism. Here, we have to mention the historical background also for attaining a proper perspective. In the beginning of 19th century, William Paley (1743-1805) suggested that we can get the idea of a creator God by observing the complexity of objects in nature. He used the analogy of a watch found lying in a deserted place. Since the watch is a complex machine we ordinarily make a logical inference that it was designed and manufactured by some intelligent person. In similar way, the apparent complexity of natural things prompts us to think that they have a creator that is God. This proposal is called *Intelligent Design Argument*; it is in accordance with the postulate that we can rationally infer the existence of God from the observations of natural world.

But Paley's line of thought – popularly called *the watchmaker argument* – was famously rejected by David Hume who argued that we get only empirical knowledge about natural world. Considering the diametrically opposite views of William Paley and David Hume, the notion of Intelligent Designer is problematic. However, recently in 1990s, a group of biologists and thinkers tried to renovate this thesis adopting the empirical approach for explaining the order and complexity in natural world. American scientists mainly Michael Behe and William Dembskey published books for promoting the Intelligent Design Argument in the context of biological evolution. They aim to give a secular and scientific flavor to the Designer so as to avoid the problem of linking religious God with science. This movement is specifically meant to attack the atheist mood created by Darwinism, which holds that biological evolution happened through materialist processes and hence, the belief in the existence of God is baseless.

An overview of Intelligent Design Argument, currently debated among biologists with religious orientation, consists of two principles as under.

- Irreducible Complexity Principle
- Anthropic Principle

Taking into account the discoveries of molecular biology and genetics we can say that cells and organs of an organism are like highly complex machines. In a complex machine we can see the mutual compatibility of various parts. If we try to remove one part and replace it by a slightly different one, then the machine will not work. In other words, the parts of a complex machine cannot be replaced one by one; to change the machine we need an entirely new design and make a new machine as a whole. This situation is termed as Irreducible Complexity Principle.

Michael Behe, William Dembskey and their friends argue that Darwin's theory of small variations (mutations) is not feasible since each part of organism has irreducible complexity. To buttress this point, they take the example that the eye of an ape cannot be changed bit by bit to form the eye of a human being. The evolution of eye has to be accompanied by corresponding changes in brain and nervous system as well as the modification in the tissue structures. The primates evolved into monkeys, chimpanzees and humans by repeating the stages of new design; that is, in every stage of evolution there is a new holistic design of a higher organism. This necessitates the existence of an external Designer with cosmic intelligence. The Designer is like an artist who draws a picture many times in order to get the satisfactory one, throwing the unsatisfactory drawings into the waste basket.

The term anthropic means "of or relating to human beings or the period of their existence on earth". Accordingly, anthropic principle supports the view that the origin and evolution of universe happened for the existence of human beings. There are two versions for this principle in cosmology: (1) Conditions that are observed in the universe must allow the observer to exist. It is called *weak anthropic principle.* (2)The universe must have properties that make inevitable the existence of intelligent life. It is called *strong anthropic principle.*

We can recognize that the foregoing treatise is the extension of the points discussed in the context of inanimate physical world. Accordingly, the belief in Intelligent Designer is an Inference to the Best Explanation (IBE). This term was introduced in connection with scientific realism. However, IBE suffers from the problem of induction and so it leads to skepticism.[# 4][*]

The concept of Intelligent Designer stands for Cosmic Mind from empirical point of view. Since the proponents adopt scientific realism about the existence of matter, they will slip into the problem of *body-mind dualism*, the most important riddle of epistemology. *In this context we can note that Intelligent Designer Argument is more or less similar to creationism, which is an unhealthy mixture of metaphysics and science.* The creationists tend to argue that their subject is a science; it can be dismissed by showing that religious faith and science are two different areas of knowledge. We can conclude that the so called creation science is a pseudo science.

3.5 The System Model of Biological Evolution [# 5][*]

So far we discussed the issues of the conflicting doctrines about Biological Evolution under process view – spiritual view (Lamarckism) and physical view (Darwinism). The main lacuna of these approaches is that they are descriptive focusing on the changing aspect of evolution. We need content view as well as process view for acquiring full understanding about an entity. The complementary nature of content view and process view must be emphasized here. Remember that in the process view we get only a description of the circumstances or factors of evolution. *Though evolution is a process happening through the passage of time, we must see it in a static way in order to find its cause. The notions of existence and cause-effect belong to the content view of knowledge.* It entails the ontological perspective on evolution. We will proceed to develop an innovative philosophy that will explain the hierarchy of species existing in biological world from the perspective of content view of knowledge. In this pursuit, system philosophy endeavors to reconcile primarily *vitalism* and *naturalism* with the caveat that we must subsequently unify the process doctrines also.

At the outset it is necessary to clarify *whether there is purpose in nature*. The everyday occurrence of opposites called development and destruction -- or growth and death -- have prompted some people to think that there is no purpose in nature. In philosophical way this point is expressed as that nature has no *teleology*. Note that the notion of teleology means good purpose and it is usually associated with the religious view of nature holding that God created nature with good purpose. The **atheists** traditionally choose to refute the religious view citing the enormity of destruction, pain and death in phenomenal world. In this way they want to prove that God does not exist. Here we do not intent to discuss the philosophical issue about the definition and existence of God. Rather, we focus on the notion of purpose in nature in order to expose the fallacy of counter arguments, especially in the context of biological evolution.

For our present deliberation, the term *purpose* needs a specific meaning in observational terms. The most visible aspect of organic evolution is the achievement of great diversity and ascending order of complexity in biological world. *Taking into account this fact we can define **purpose** as the tendency of organisms to evolve to higher level of complexity.* But the advocates of materialist and atheist doctrines do not agree with the linking of evolution to the idea of purpose; their position may be criticized as following.

For arguing that there is no purpose in evolutionary history, the concerned group of materialists and atheists point to the massive destruction of intermediary species in the process of natural selection. It is estimated that the number of species which perished in the struggle for existence is a huge majority as compared to the small number of new species survived upon earth. It is similar to the death of most of the soldiers for winning a war. Such death and destruction can be treated as evil and we are disturbed about the existence of evil. Philosophically speaking, this predicament is characterized by the phrase *problem of evil*. The atheists often ask: can there be so much evil and destruction in the evolutionary process if there is a God? A perfect and all-loving God would cause the favorable mutation only so as to lead to a new species straight away without involving the production of imperfect species and their eventual destruction. Moreover, the all-powerful God need not make evolution as a complex affair. As per this line of thinking, the atheist biologists mainly Richard Dawkins, prefer to accept the materialist notion of *chance* instead of divine guidance.

Now we can clarify that purpose has both positive and negative aspects since it is a value. The positive aspect of the evolutionary process results in the formation of new species with higher complexity while the negative aspect constitutes the destruction of intermediate (unsuccessful) species as well as the imperfections in surviving organisms. This proposal will enable us to make a secular and scientific interpretation about the good and evil aspects of evolution. As per the foregoing, it must be admitted that there are positive and negative (good and evil) directions to the purpose in evolutionary process. The opposite directions of purpose are implied when we define *evolution* as the progressive change in some organisms and its future generations pertaining to a species, resulting in the emergence of a higher species.

The system philosophy of life introduced earlier clears the way for synthesizing the four different worldviews about life – vitalism, mysticism, naturalism and physical process view. We have shown that though macromolecules (DNA, RNA and protein) and information do not have separate existence, these are opposites that make the dual parts of a whole called cell or organism. **The System Model of Biological World** introduced therein serves to describe the process of biological evolution having the hierarchy of organisms from bacteria to humans. We represent the ascending levels of organisms using the graphs called *isoquants*. Accordingly, a particular species is represented by an *isoquant*, in which the different points denote the countless number of concerned individuals. [# 6][*].

Thus the lowest isoquant stands for the various kinds of bacteria which are organisms with single cell. Large varieties of bacteria are formed by the mutation of some genes in the first cell and subsequent division of cell as well as by the repetition of the process innumerable times. In the next stage, the joining of various groups of bacteria and genetic mutation resulted in the formation of multicellular microorganisms;

these species are represented by higher isoquants. The repetition of this process through evolutionary history has resulted in higher order species like plants, fish, reptiles, birds, animals and finally humans. All these species can be represented by a sequence of isoquants in the X-Y model; this diagram is now called ***The System Model of Biological Evolution***.

This model is a physical representation of the hierarchy of organisms. However, it may be emphasized that we accept three axiomatic principles as below.

- Evolution has both physical and nonphysical aspects.
- Content view is necessary to know the cause and ontology of evolution.
- Each worldly phenomenon has the dual features of permanence and change.

Since *information* is the physical version of genetic code, the first principle holds that each species is a dialectical combination of macromolecule and genetic code, in the form of X and Y axes respectively. That is, every organism has physical and nonphysical parts. It may be clarified that here genetic code is responsible for the purpose, creativity, design and planning that are observable in the stages of biological evolution.

The second principle asserts that cause-effect is a notion pertaining to the content view of knowledge, which is the ontological perspective. The system of macromolecule and information (X-Y system) has phenomenal existence and it functions as the cause of evolution. This system is like an artist who wants to draw the best picture. The artist throws the unsatisfactory drawings to the waste basket and repeats his creative work till he reaches the desired level of perfection. This analogy can be used to understand the discrete levels of species in the fossil collection. Since the isoquants are separate graphs, there is a clear gap between two consecutive species. It means that the evolution from one species to a higher species involve visible gap and jump, when we adopt content view. The mutation of interrelated genes of a species and natural selection altogether is a creative event which results in the formation of a higher species.

The third axiomatic principle urges us to study evolution under content view in addition to the traditional method of process view. We have defined evolution as the change to a higher level of complexity. But this phenomenon has both the aspects of permanence and change. When we observe the hierarchical levels of organisms, we are concerned with the aspect of permanence for a finite duration. A particular species that exists today may become extinct after a period. The formation of various levels of species in the evolutionary history is a matter pertaining to temporal permanence. On the other hand, the emergence of a particular new species from a preexisting lower species is a topic to be considered under process view. The mainstream of evolutionary science including Darwinism focuses on this aspect of change. The complementarity of permanence and change is to be born in mind for an enlightening study of evolution. [*].

Life and Mind

Chapter 3/Diagram
System Model of Biological Evolution

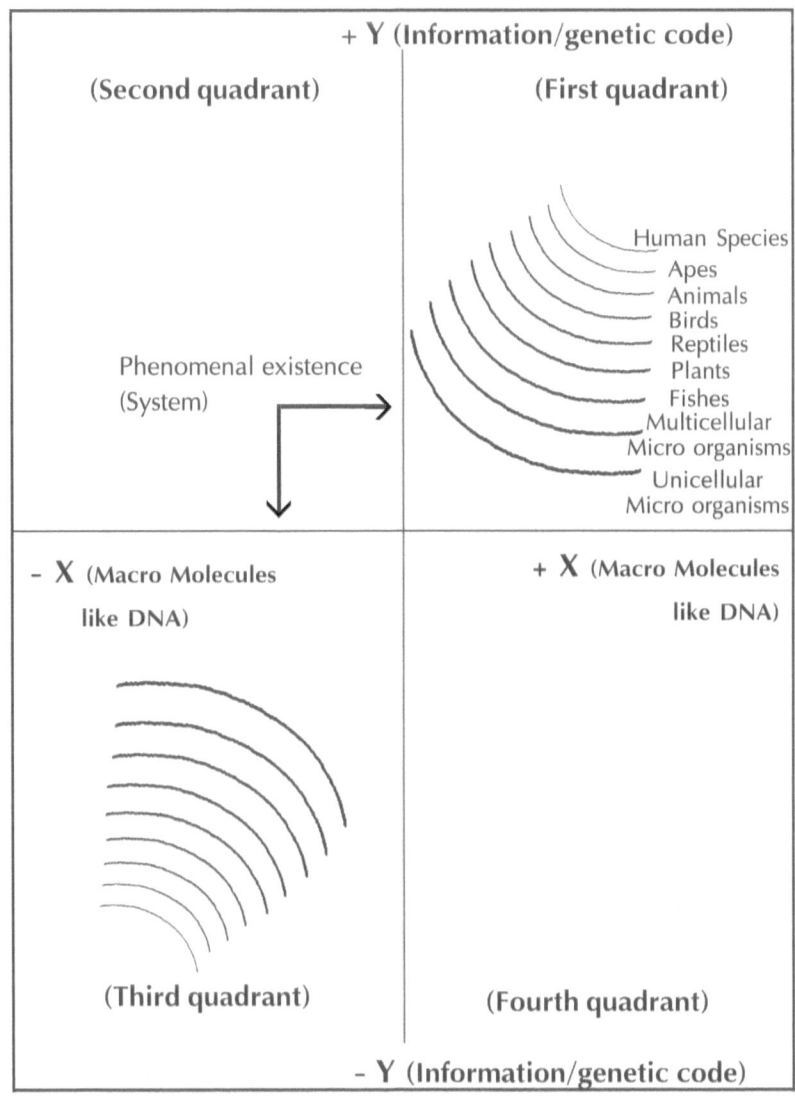

The foregoing principles of system model would equip us to tackle the *problem of missing links* in an ingenious way. When the scientists arrange the collected fossils of species in an ascending order, they find that there are gaps between two consecutive species. In this situation, scientists speculate that there are some intermediary species belonging to the gaps, about which fossil evidences are not yet available. Such species are called *missing links* in the fossil line. A problem arises here when we ask: do the missing links indicate the creative jumps in evolutionary history so that Darwin's theory about small variations is to be refuted? The heated debates between creationists and evolutionary scientists regarding this issue can be resolved now as per the following paragraphs.

Richard Dawkins in his book *The Greatest Show on Earth – the Evidences for Evolution* (2009) argues that the notion of missing links is superfluous in the context of scientific study of evolution. There may be many unknown intermediary species in the process of natural selection. It is irrelevant to debate upon such species about which fossil evidences have not been obtained. The principle of natural selection explains coherently the evolution of a higher species from a lower one, in spite of the non-availability of fossils about intermediate species. Through complicated lines of arguments Dawkins becomes the most vociferous follower of physical process view contained in Neo-Darwinism. However, since Dawkins' basic premise can be refuted by our philosophical analysis, there is no significance to his treatment of gaps in fossil line. [# 7].

The problem of missing links is a favorite topic for the advocates of the religious view of creation and fixity of species. Their fundamentalist argument is that the gaps in the fossil line imply the stages of divine creation. The *intelligent design argument* is another version of the religious position, adopting the empirical perspective. We have already rebutted the religious view of creation since it involves the controversies about the existence of God. The problem of missing links essentially arises when the creationists bring in God to fill the gaps of fossil sequence. We may reiterate that, for interpreting the missing links, the physical process view about natural selection or the religious theory of creation cannot be admitted on account of the reasons just highlighted. In this situation, System Philosophy can innovatively interpret the idea of missing links as below.

Suppose that A and B are two consecutive species in the fossil record, the natural selection from A to B happened over many millions of years in accordance with the system model. It is reasonable to conceive about many intermediate species between A and B. But the non-availability of fossil evidences may be due to geological reasons or other factors. In spite of this fact, the principle of natural selection is valid for explaining the evolution from A to B. Now assume that, in future, scientists get the fossil of an intermediate species; let it be denoted by A1 for placing in the fossil line between A and B. In this situation, the principle of natural selection holds good to explain the evolution both from A to A1 and from A1 to B.

The *System Model of Biological Evolution* under content view accepts the notion of missing links. It represents the gap and jump between two consecutive isoquants, caused by the dialectical productive relation between physical and nonphysical aspects of biological evolution. *Since the notion of missing links is a static view about the levels of species, it cannot be incorporated in the process view of evolution.* Now we may emphasize that the process scientists like Richard Dawkins are concocting frivolous arguments against an idea that lies outside their methodical framework. The acrimonious debate between evolutionary scientists and creationists is the result of the ideological differences between materialism and theism; we find it as irrelevant in our scheme of analysis. *The integrative approach of system model eliminates the said controversy by limiting the idea of missing links exclusively to the content view.* The advantage of system model of evolution is in illustrating the existence of a hierarchy of species without bothering about the fact of missing links. [# 8].

Next we will note that the above Diagram can be used under process view also. In this situation, it is beneficial to adopt the concepts originally introduced by Charles Darwin, albeit by suitable adaptation. Accordingly, the formation of a new species from an ancestral species happens on account of four factors namely mutation, struggle for life, inheritance and natural selection. **System Philosophy modifies Darwin's theory by holding that each of the four factors is a combination of macromolecule and information.** Our method incorporates the material and creative aspects pertaining to mutation and subsequent factors of evolution. In this way the same diagram serves to depict the various levels of species under the process view as well as content view. The integrative method presented above envisages that there is a sort of mystery in evolution and it is beyond the frame work of science. The creativity and purpose inherent in evolution is comprehensively explained by the system model of macromolecule-information. In this scheme of thought, there is no place for the notion of creator God or intelligent designer, which suffers from the issues of realism.

For completing this philosophical deliberation, we have to connect biological evolution with the development of inanimate world and here we rely on the notion of *fine tuning* introduced earlier. The evolution happened to astronomical bodies, especially earth and its atmosphere, during the long ten billion years after big bang was a sort of preparation for the emergence of life. This evolutionary process in the inanimate world can be logically construed as displaying the nonphysical aspects of purpose and creativity in addition to material complexity. From this point of view, the sequence of developments in biological and inanimate fields can be unified as the phenomenal effect of a deeper reality. We can conceive the ultimate reality of universe as **matter-consciousness system**; it is the essence of the *theory of reality* under system philosophy, to be technically articulated in the next book.

Then what is the future of Darwinism? Admittedly, it is a pragmatic and scientific theory only for explaining organic evolution through physical process view. The nonphysical aspect of evolution – creativity, purpose and freedom – is outside the scope of science exclusively constructed in physical terms. According to this method, Darwinism attempted to translate the nonphysical aspect of evolution into the physical processes of DNA and other macromolecules of organism. In this manner, the genetic code is treated as an algorithm, alternatively termed as information, representing the mechanical activity of macromolecules. This computer model is applied to describe physically the factors of evolution, namely, mutation, struggle of life, heredity and natural selection.

We have mentioned earlier that the scientific method has helped biologists to collect material evidences – fossils and suitable geological data – for deriving inferences about the stages of evolution. The so called *tree of life* is constructed on the basis of such physical evidences; it is the practical achievement of evolutionary scientists. *But they unwittingly entertain the fallacy of treating biological evolution as a physical process.* Moreover, the justification of theories and inductive laws pertaining to various physical phenomena constitute an unsettled issue in philosophy of science. In spite of the baggage of epistemological problems, the protagonists of Darwinism would continue their research program for practical and ideological reasons.

We may conclude this chapter by adding a few points as a final review of the **atheist** writings of Richard Dawkins. The overall objective of his books is to counter the creationists by advancing the project of Darwinism. He hammers on the point that, since the physical process approach is strongly supported by scientific evidences, creationism is refuted by default. *But we can comment here:* the strategy of positioning the scientific view against the religious view reflects the age old conflict between materialism and idealism. Both these philosophical theories involve the drawbacks of realism; It is the springboard of the antagonism between science and religion in the context of biological evolution. To be more specific philosophically, one side consists of scientism and atheism while the opposite side is formed by creationism and theism. Each side is engaged in shadow boxing. We have cleared this mess and erected the edifice of system philosophy on firm grounds for a comprehensive and integrative knowledge about evolution.

NOTES of Chapter 3

\# 1   Darwin (1859), *On The Origin of Species* (Dover Edition, New York, 2006); Darwin (1877), T*he Descent of Man* (Second edition in 2004 by Penguin Classics, London)

\# 2   Dawkins (1976), *The Selfish Gene* (Oxford University Press, Oxford and New York, 1976)

| # 3 | The foregoing epistemological analysis of Darwinism is my original idea. Moreover, it enables us to apply System Philosophy to the aspect of creative evolution, by refuting the approach of spiritual or mystical view of biological evolution. |

| # 4 | The concept of IBE has been explained in the previous *book, Origin of Universe* |

| # 5 | For record, it may be stated that *The System Model of Biological Evolution* is my original idea. |

| # 6 | The concept of isoquant has been borrowed from the production function model of economics. It is the specialty of *The System Model of Biological Evolution*. |

| # 7 | Dawkins (2009), *The Greatest Show On Earth: Evidence for Evolution* (Bantam Press, 2009). |

| # 8 | The application of System Model to interpret missing links would refute both creationism and atheism. |

CHAPTER 4

SYSTEM PHILOSOPHY OF MIND

****

4.1 Human Mind and nervous system

4.2 Conflicting Philosophical View about Human Mind

4.3 System Philosophy of Mind

    4.3.1 Structure of Human Mind: the System Model

    4.3.2 Solution of Body-Mind Dualism

    4.3.3 What is Consciousness?

Author's main original ideas are marked by [*]

The mark [#] gives the number of note at the end.

****

Higher organisms, especially human beings, have three levels of structure as physical body, life and mind. This is as per the content view of knowledge about organisms. The term biological body is used to refer to the combination of physical body and life. Biological body consists of cells, tissues, and organs in a hierarchical order; all these parts have specific biological functions. But the special organs such as brain and sensory organs are together called *nervous system,* which additionally perform certain mental functions such as feeling, willing and thinking. The totality of mental activities is conventionally termed as **mind**. Accordingly, we hold that ***mind is a higher phenomenon which exists over and above the biological processes of nervous system.*** Here, we distinguish between the functions of life and mind.

In this context, we can treat **mind** as a factory which produces various mental phenomena like emotions, desires, thoughts and memories and at the same time as a being which controls various activities of biological body. We are most aware of our own mind, but are doubtful whether mind exists in lower beings like plants, birds and animals. The difference in the mental capabilities from person to person is highly baffling. At the same time, we often wonder how human species acquired the complex form of mind in the course of biological evolution. The religious interpretation about the phenomenon of mind adds to the mystery in this field of enquiry.

We may tentatively accept that ***philosophy of mind*** aims to explain the deeper questions about the existence of human mind and its activities. The foremost issue in *philosophy of mind* is the definition of mind since we have to take into account the related notions like body, soul, spirit and consciousness. There are great differences between science and religion while considering the question: what is mind? More specifically, the development of psychology as the scientific study of mind is at loggerheads with the religious approach. It is necessary to reconcile such conflicts by adopting a logical and secular point of view. We can anticipate that the system model of life and evolution presented earlier would show the path for dealing with the next level that is mind. This approach will lead to deeper insight about the ultimate reality behind the inanimate and living things of this universe. [# 1].

The central theme in the course of this chapter springs from the age-old problem of distinguishing between mind and soul. We will explain that, in the philosophy of Rene Descartes, mind is identified as the metaphysical being in tune with the religious notion of soul. He introduced the issue of body-mind dualism, which remains as the greatest riddle in the entire field of philosophy. [#2]

In order to make headway, the present chapter deliberates upon the traditional dilemmas *regarding mind, as existing in human being,* because these are relevant in the philosophy of science also. The latest advancements in scientific study about mind – various branches of neuroscience and psychology - constitute the ground from which we will start our discussion.

4.1 Human Mind and Nervous System

From the view of biological sciences, the nervous system of a human being has the following parts.

1. Central nervous system
    a. Brain
    b. Spinal cord
2. Peripheral nervous system
    a. Cranial nerves or somatic nervous system
    b. Autonomous nervous system

It may be noted that all parts of nervous system are basically made of a particular kind of cell called *neuron*, which has the capability to transform the impressions of light, sound, heat and cold into electric signals. Such signals pass through various nerves, where each nerve is formed by connecting nerve cells like a chain. Brain is the central part of nervous system since the main nerves start from it. Certain chemicals called neurotransmitters are produced between two adjacent neuron cells for causing the flow of impulses or signals through various parts of nervous system.

We have five sensory organs called skin, eye, ear, nose and tongue which produce different kinds of signals (or *impressions*) which are transmitted to the brain through specialized nerves. When such signals reach the sensory parts of the brain, certain neural networks and mental images are formed; this stage is called *sensation*. The internal events of reflection, including the activation of certain parts of memory, also produce sensation in brain. Subsequently, certain processing of sensations takes place by forming suitable *organized neural networks* and *ideas*. We can simply say at present that these neural networks serve as the foundation for our mental states like emotions, feelings, desires and thoughts. In other words, **the phenomena of mind occur and exist above the networks of neuron cells in the brain.**

A specialized branch of biology termed *neuroscience* has been developed in the recent decades for observing and analyzing the functions of brain and other parts of nervous system. Bacteria, plants and some other organisms do not have specialized nerve cells – so we can say that the mental function is embedded in the genes of general cells. *Taking into account of this fact, we accept the proposition that mind exists only in animals and other organisms which have well developed nervous system, especially brain.* **The ensuing parts of this chapter deal with the matured form of human mind only.**

It is expedient to mention here, though tentatively, about *consciousness* that is the essential property of mind. We can list the main mental events with the property of consciousness as perception, thinking, learning, attention, motivation and emotional experiences. Note that the word **consciousness** is normally used in third-person perspective. But when a person becomes aware of a mental state in first-person perspective, the person is said to have *self-consciousness* about the mental state.

In order to study our mental events in an integrative manner, it is necessary to hold that mind exists in two states namely, **conscious mind** and **unconscious mind**. The lower part of brain is responsible for unconscious mind, while the upper part of brain – including cerebrum and cerebral cortex – causes conscious mind.

The various organs in the lower part of brain are medulla, ponce, cerebellum, thalamus, limping system and hypothalamus. They function to produce the *unconscious mental states* such as desires, emotions, feelings, motivations, memory and instincts. Additionally, they control various biological activities like respiration, blood circulation, sweating, chewing, metabolism and excretion. For this purpose, the signals originating from lower brain flow into the cells of concerned organs – lungs, heart, intestines, skin and so on. After entering the cells, such signals cause the expression of particular genes controlling the production of proteins responsible for the bodily functions. The mental activities and bodily movements related to **unconscious mind** are *involuntary* because they happen without the knowledge and control of conscious mind. When we talk about mental health we are focusing on the harmonious function of various organs of brain just mentioned. It implies that the various mental diseases such as depression, anxiety, phobia, schizophrenia and epilepsy are manifestations of the irregularities in the lower part of brain; it accounts for the derailment of unconscious mind. [# 4]

We can arrange the activities of **conscious mind** into three categories as following. [*]

1. *Subjective experiences*
    - *Imagination* – hallucinations, dreams and literary imaginations
    - *Emotions* – pain, fear, anger and happiness
    - *Conations* – desire, purpose and motivation

The subjective experiences or feelings can be converted into propositions in first person perspective. Examples: I had a dream last night, I am afraid of snakes. These sentences express the subjective experiences using the concepts of language and grammar.

2. *Objective propositions about experiences*

It is a special power of self-consciousness to convert the subjective experiences into ideas and propositions of third-person perspective. In this manner, for example, I have objective ideas about my pain resulting in a proposition like "my pain in stomach is not serious".

3. *Knowledge about external objects*

These are objective ideas and propositions produced in the conscious mind on the basis of sensory experiences, logical thinking and application of language. Most of our knowledge about external objects belongs to this category. More specifically, we can say that the objective knowledge is formed by various conscious mental activities like comparison, mathematical calculations, drawing inferences and combining various ideas. In the case of mathematics and other abstract subjects, the role of sense experiences is less important as compared to the creative aspect of thinking.

From the above classification, it can be inferred that the conscious mind primarily generates subjective experiences and propositions. The defining property of such experiences and ideas is *self-consciousness*, as mentioned earlier. But the predominant function of conscious mind is the production of objective ideas in third-person perspective about external objects as well as our own experiences. When I am writing these lines I have the objective knowledge that I am thinking such and such sentences.

The crucial point here is that the mental states produced by mind have the wonderful property of consciousness. *We define consciousness as the essential property of a mental state so that the concerned person is aware of the occurrence of such mental state in first-person perspective.* For example, pain is a mental state with the property of consciousness because I become aware of the occurrence of pain. So I am able to say: "I have pain". Consciousness refers to the ability of a person to know the external world as well as the person's own mental states.

As per our analysis, the self-consciousness has three characteristics as given below.

Nonphysical aspect: The neural network occurring at a particular part of brain and nervous system can give rise to a large variety of mental states. It leads us to the suggestion that a specific mental state does not have a physical location. We can add that mental states are nonphysical though they supervene the physical structure of neural networks.

Content: The idea contained in a sentence is called content. Here, we have to distinguish between idea and language. It is possible to say that different persons speaking English and Hindi respectively have the same idea about the thing used for writing; it is called pen in English and *kalam* in Hindi. The common idea is alternatively termed as content. We can add that content is the defining property of ideas and propositions in third-person perspective. In the case of a particular emotion or feeling also, two persons have the same content though they may express it in different languages.

Intentionality: The intimate relation between conscious mind and unconscious mind may be considered now. As explained above, various kinds of experiences and knowledge are produced by conscious mind. But unconscious mind is the repository of biological instincts, habits and complexes – they represent the various purposes of life. When I see a knife, I can treat it as a weapon or as an implement for cutting vegetables. The purpose inherent in unconscious mind in the context of formation of ideas and propositions is called *intentionality* and it is the most basic aspect of self-consciousness. Various kinds of superstitions, myths and ideologies are precipitated in our unconscious mind and they contribute significantly to our knowing process.

Since ideas are prior to linguistic form, we can compare language to a bottle for packaging. The ideas formed in unconscious mind are packed in the bottle of language to become thought for expressing in oral or written form. *(It may be remarked that in ordinary usage we treat idea and thought as synonyms).* In this context, we can underline that science is a particular combination of idea and language which is to be contrasted with that of religion or art. The streams of mental states occurring in human mind are organized in various stages to constitute the spectrum of knowledge.

As per scientific perspective, it was an evolutionary leap when humans started to invent symbols, words and higher forms of language in order to express their thoughts and feelings. The term *cognitive mind* is used to denote the higher faculty of mind for producing knowledge; it involves the processing of sensory experience as well as the application of logical reasoning and language. Based on the considerations of theory of knowledge, it is now possible to propose that our knowing mind consists of the dual parts namely *intellectual mind* and *mystic mind*.

The former tries to understand things of world using sensory data and rational thought; it has two parts -- lower part is the *philosophic mind* and the upper part is called *scientific mind*. On the other hand, mystic mind produces the experiences and knowledge about religion as well as art. Further, both conscious mind and unconscious mind have these divisions. Then the structure of cognitive mind as a whole can be depicted by the following diagram.

Table 1 : Structure of Cognitive Mind [# 6][*]

	Intellectual mind (*philosophic mind* and *scientific mind*)	**Mystic mind** *(religion and art)*
Conscious mind	Intellectual conscious mind (C)	Mystic conscious mind (D)
Unconscious mind	Intellectual unconscious mind (A)	Mystic unconscious mind (B)

The systematic study of the aspects of human mind becomes more complicated in the wake of the above table. How can we explain the said structure of cognitive faculty? It is now necessary to focus on **the crucial question: what is the power or agency that produces the different kinds of experiences and knowledge?** The subject of *philosophy of mind* aims to answer this question considering self-consciousness as the distinguishing aspect of human mind. [# 5].

4.2 Conflicting Philosophical Views about Human Mind

In this context we may recall the worldviews, which serve as the foundations of the main areas of our knowledge. The worldviews can be classified under content view and process view, each of which is further divided into rationalism and empiricism. Accordingly, there is a 2X2 table for the different theories about mind as shown in the Table 2 given below.

Table 2 : The Worldviews about Mind [# 7][*]

	Content view	Process view
Rationalism	1. Organic world view and mechanistic world view-rational (Idealism or Dualism)	2. Spiritual process view – rational (parallelism and panpsychism)
Empiricism	3. Mechanistic world view- empirical (Naturalism *or* Epiphenomenalism)	4. Spiritual process view – empirical (phenomenology) 5. Physical process view (computer model functionalism and neuroscience)

The idealist philosophy of mind is about soul and its manifestations called life and mind. Descartes (1596-1650) is the most famous exponent of this metaphysical view. His doctrine implies a **dualism** between soul and physical body in the case of a human being. As such the soul is a metaphysical being with the power of thinking rational ideas. Soul originates from God and comes to reside in the physical body during the period between birth and death. The interaction between soul and physical body cause the phenomenal aspects of life and mind.

An important aspect of the Cartesian view is the clubbing of sensory experiences – emotions, desires etc. – with biological functions of body. It may be remarked that this notion is the legacy of the ancient philosophy of Plato. Further, such conception existed in the Indian philosophy of mind also. Accordingly, the **chariot model of mind** became popular by which the rational soul is like the driver or charioteer. Plato described the rational soul as a *homunculus*, which means *little man*, inside the head. The existence of this inner person raises serious philosophical problems when we ask: how did homunculus get this power of reason? To answer this question, we must say that there is a smaller inner person inside the homunculus. But the question can be repeated *ad infinitum*, this dilemma is known as *homunculus fallacy*. It is necessary to take into account this fallacy also when we consider the floodgates of criticism on the body-mind dualism of Descartes. We may summarize it as below [# 8].

a) Descartes treated animals other than humans as automata having only biological bodies to be studied by science. He did not recognize any sort of mind in animals. This position does not agree with the tenets of psychology that establishes the hierarchical order of mind in all living beings. The crucial question in this situation is: how can we account for the exceptional qualities of human mind as compared to the lower levels of mind in other organisms?

b) The interaction between body and mind is observed in our voluntary actions as well as in the biological functions. For explaining this fact, Descartes advocated the metaphysical theory called *interactionism*. But he failed to explain how two ontologically different substances – matter and soul – can come in contact. He brought the notion of creator God

into this context; it attracted serious difficulties from philosophical angle.
c) There is no satisfactory solution to *homunculus fallacy* as long as we adopt metaphysical realism. The existence of rational soul as a being is a persisting issue in philosophy of mind because we have to reconcile between the separate meanings of soul and mind. This has deeper implications with regard to the science-religion conflict.

The metaphysical articulation of body-mind dualism is challenged later by the critical philosophy of Immanuel Kant. Accordingly, there is no justification in holding that matter and soul are real substances because we have only phenomenal knowledge about the constitution of our being. Kant argued that metaphysics is impossible in the context of philosophy of mind as well as scientific study of world. This leads us to the dilemma: what is the phenomenal structure of human mind and what is its reality? [#9][*].

The **spiritual process** doctrines of mind are parallelism, panpsychism and phenomenology, which would come under the broad philosophy of process. Spinoza (1632-1677) and Leibniz (1646-1716) appear as the chief philosophers to be considered here. They famously advocated the doctrine called *parallelism* which advocates that body and mind are parallel processes existing in a human being. To express this view as per modern terminology: when the neuron network is formed in the nervous system, corresponding mental states occur in a harmonious way. The parallel existence of neuron networks and mental states is explained by holding that immanent God (Spirit) is responsible for this arrangement. This view is articulated as part of a doctrine of mind called *panpsychism,* holding that all inanimate things as well as living organisms found in plant and animal kingdoms possess the mental power with varying levels of complexity. The implication is that all things of nature have a mental aspect as well as a physical aspect. [#10].

Now we may focus on the recent doctrines under spiritual process view of mind. Typically, such doctrines also hold that the worldly phenomena of inanimate things and living beings are manifestations of immanent spirit (God) working inside the material bodies. Many thinkers, including a few scientists, with religious orientation have attempted to explain mind along this way. The pioneering philosopher of this group is Teilhard de Chardin (1881-1955); his masterpiece is *The Phenomenon of Man* written in 1938, but published posthumously in 1959. [#11].

In the recent decades, there is a new mystic version of mind known as *quantum consciousness* (sometimes called *quantum mind*). It expounds that consciousness is associated with the quantum processes happening in the subatomic level of neuron cells constituting the nervous system; the uncertainty involved in the quantum processes gives rise to the consciousness. There is a long array of neuroscientists and fellow researchers who try to interpret consciousness resorting to the quantum fields occurring in the neuron cells of *brain and other parts of nervous system (BNS)*. [#12].

Now it may be reiterated that Descartes was wrong in invoking the dualism in ontological sense. The failure of Cartesian philosophy eventually paved the way for the development of materialist doctrines of mind together with experimental methods of psychology. The materialist theory of mind is called **epiphenomenalism,** as introduced earlier. It suggests that we can explain the conscious mental activities by appealing to material processes happening in brain. There are two main purposes behind this approach. Firstly, it tries to tide over the philosophical difficulty pertaining to Cartesian *dualism* and *interactionism*. Secondly, neuroscientists could utilize this theory to strengthen the study of brain through experimental methods. They want to detect the physical processes in brain so as to explain the conscious mental states in a practical manner. It amounts to the view of *eliminativism*, which says that the mental properties - especially the consciousness – could be eliminated from the purview of scientific investigation of mind. [# 13].

Though epiphenomenalism historically promoted the experimental methods of classical psychology, this approach could not explain the creativity and freedom of our thoughts as well as the motivations for our actions. In this period of confusion, another group of psychologists could recognize that mind has a subconscious level which significantly determines our actions and other external aspects of mind. Then, psychology and philosophy of mind were subjected to big churning in order to promote an evolutionary worldview.

Charles Darwin presented the theory of biological evolution through the book *On The Origin of Species* (1859). Extending this principle to the realm of mind, William James in 1890 proposed that mind is an activity of organism for achieving the goals of adaptation and survival in adverse circumstances. He called his philosophy of mind as *functionalism*. It emphasized the principle that the neuron networks are formed in the brain of an individual on the basis of the goal for adapting to the environment. As per this approach, every activity of stimulus and response is interpreted as a strategy for survival.

The terms like instinct, motivation, urge and desire gained predominance in the new evolutionary conception of mind and it projected the function of unconscious mind to the realm of experimental psychology. Franz Brentano (1838-1917), Sigmund Freud (1856-1939) and Carl Jung (1875-1961) are the most famous psychologists who pioneered the scientific study of *unconscious* states of mind. They developed a new branch of psychology called *psychoanalysis* which focuses on the tensions and conflicts in the unconscious mind in order to discern its effect on conscious mind including various kinds of actions.

Great advancements were made during 20th century in various branches of psychology by adopting the functional approach. Consequently, the mental states are treated as activities meant for achieving certain goals of individual organism. Of course this is an interpretative and descriptive approach for dealing with mind as a process. Adopting this path, the philosophy of mind took a turn to the new paradigm termed as **physical process view**. We can find three doctrines proposed in this area – identity theory, behaviorism and computer model functionalism (CMF) – to be explained briefly in the following paragraphs. [# 14] [*]

The development of neuroscience is the background for proposing the **identity theory**. With the help of advanced imaging technology, neuroscientists are able to link certain mental states with the stimulations happening to the neuron networks at particular parts of brain. Since neuroscientists try to interpret mental functions in terms of stimulations occurring in corresponding parts of brain, they adopt the view that a particular mental state is identical to a particular brain state. In other words, the brain state conveys the same meaning that we normally attribute to a mental state. The application of this scientific approach is limited to the treatment of certain diseases. But it is incapable of accounting for higher mental functions like learning, designing and calculating which involve the coordinated function of various groups of neuron networks.

The pioneer of **Behaviorism** is the American psychologist John Watson (1878-1958), who presented this idea in a thesis in 1913. According to him, psychological experiments follow the theory of knowledge called logical positivism in which certain unobservable theoretical entities are given meaning through experimental evidences. In the context of behavioural psychology, for example, the mental state called *fear* is an unobservable theoretical entity. But we can make a hypothesis in observational terms that fear is expressed as particular set of expressions in face as well as other bodily parts. Such a set of expressions, conceived as a stimulus-response relation, is called *behavior*. Suppose that a person sees a venomous snake. This sight is a stimulus that causes a set of expressions like surprise, making sound and running away, which together is called response. There is no need of saying that fear has ontological existence; it can be treated as an activity observed as a particular stimulus-response relation. In this way the theoretical term *fear* gets meaning.

We can criticize Behaviorism by pointing out that the verification principle of logical positivism is not applicable in creative forms of mental states like designing, calculating and willing. In such cases, we cannot know the true kind of mental state by observing behavior. Hence it is not valid to say that a particular mental state gets meaning by the evidence of external stimulus-response relation. That is, the scheme of behaviorism is not satisfactory, leading to its decline and fall in 1950s. The contemporary advancements in computer science opened a new way for psychological study by comparing the function of mind to the activity of a computer. It was recognized that the *central processing unit (CPU)* of computer works like an *input-output* device. This information processing model of computer provided new insight to psychologists that behaviorism could be modified drastically.

It was proposed that, similar to the CPU of computer, human mind has certain *internal structures* which mediates the stimulus and response. The discoveries of neuroscience about the formation of neural networks in brain gave concrete evidence in this regard. *It was recognized that various mental states occur in association with the pattern of neural networks appearing in specific parts of brain.* Recall that a computer appears as a purposive machine, as it is controlled by the software (program or algorithm). Similarly, psychologists envisaged that mind is an algorithm that determines the emergence of suitable neural networks for performing the task of information processing. Obviously, it is a descriptive method for studying the activities of mind; it does not deal with the existence of mind in a static manner that pertains to content view. The 1960s witnessed the beginning of the era of **cognitive psychology** that utilizes neuroscience to describe the mental processes required for generating knowledge.

How do cognitive psychologists take into account the aspect of **consciousness** inherent in the mental states? Adopting the empirical approach of science, *consciousness is defined as the awareness of environmental stimuli as well as internal events like attention, feelings, thoughts and explicit memory.* This is a process view with pragmatic approach about consciousness. In this way, the conscious mental events occur when the person is awake. In contrast, a person under sleep or amnesia has exclusively unconscious mental states. So the distinction between wakefulness and sleep is the key to mark the various states of consciousness. In other words, the cognitive psychology proposes a direct correspondence between physiological measures of brain activity and the levels of consciousness.

Generally speaking, we can note that the cognitive psychology was developed utilizing the principles of computer science as well as the methods of neuroscience. This led to the articulation of a new theory about mind called **computer model functionalism (CMF)**. It is based on the observation of neuroscientists that external stimuli produce impressions in unconscious mind and it causes certain conscious mental states. This complex activity takes place over and above the networks of neuron cells. It is possible to interpret the mental processes using the analogy of a computer which has two parts namely, hardware and software.

As per the new doctrine, a particular algorithm represents a mental state like fear, idea, desire or willing; each algorithm corresponds to a pattern of neural networks. Then human mind as a whole is a super algorithm consisting of innumerable sub algorithms. *Since a specific mental state has consciousness, with the attributes of content and intentionality, the proponents of CMF argues that the concerned algorithm also has such properties.* Since an algorithm is a plan of activity, it is reasonable to admit that it is intentional. Next step is to show that the mental algorithm has representational content also.

Generalizing the above method, the followers of CMF hold that we can explain conscious mental states in terms of algorithms which have the properties of *content and intentionality*. It amounts to proposing that a mental algorithm represents particular mental states and consequently external objects – this view is called **representationalism.** Hence the framework for describing various mental states is alternatively termed as *computational / representational theory of thought* or *CRTT*. The algorithmic representation of conscious mental events is also referred to by the phrase "language of thought", because it is held as common to all human beings. [# 15]. The words and propositions used for describing various mental states are according to our linguistic practice. In this manner, computer model functionalism treats mental states as conventional terms for the description of software/algorithm pertaining to interrelated neural networks. [# 16][*].

The computer model is a practical and interpretative approach for describing the production of mental states. When we say that a certain set of neural networks corresponds to a particular mental state, the pattern does not exist as a thing. To clarify this point, consider the fact that computer software is not a thing like paper or pen; it is an organization of nonphysical ideas occurring in the mind of a programmer. The extension of this analogy to human mind creates serious conceptual problems. We can ask: who is the agent for producing the algorithm (program) of neural networks? The computer model functionalism does not consider satisfactorily this question about agent, since it adheres to the process view for description of activities.

We may reconsider the question about the origin of mental algorithm. The materialist philosophy of mind is called *epiphenomenalism*, as explained earlier, holding that the mental states are byproducts of the physical process in brain and other parts of nervous system. **Since computer model functionalism is the process version of materialism, we can say that it moves towards epiphenomenalism in order to explain the formation of various mental states. The philosophical issues of this theory have been mentioned earlier.** If mental states are the byproducts of physical processes, the concerned algorithm is mechanical. It cannot account for the mental causation that produces the motivation to do purposive actions. Nonphysical consciousness is outside the ambit of physical algorithm.

We have just established that the computer model functionalism as well as other doctrines under metaphysical, spiritual and materialist worldviews has failed in the following key questions: What is mind? What is consciousness? How do mental states occur over and above the physical processes of brain? It is the project of System Philosophy to reconcile the conflicting doctrines and present a comprehensive view about the existence of mind.

4.3 System Philosophy of Mind [# 17][*]

We have already presented the critical comments against the dogma of biologists that the mental states occur continuously due to the neural networks formed in BNS. So the scientific view envisages the linear causality from BNS to mental states, where the former is treated as physical since it is basically made of material compounds while the latter is regarded as a byproduct of physical processes. *But we have already shown that the one way linear causality from BNS to mental states is not sufficient because we observe the reverse causality also from mental states to various bodily organs. Considering the creativity, purpose and freedom of our actions, the agency role of mind is of great importance. This implies that the materialist theory of epiphenomenalism cannot be treated as correct.*

To cite an example, consider two persons walking through a beautiful garden. Suitable neural networks are formed in the particular part of their brains for representing the details of the garden. The first person simply has the ideas about the beauty of flowers while the second person experiences a flight of imagination in order to produce a poem. It is clear that the neural networks formed in the same part of BNS have generated two different kinds of mental states – simple ideas of beauty and imaginative ideas of poem – in those individuals. Here, we can conceive the aspect of creativity, which varies from person to person, with respect to the formation of neural networks in brain. In this situation, it is reasonable to doubt the validity of the definition of mind advanced by empirical methods of biology. We can accept that the nonphysical mind cannot emerge from physical BNS; the scientist view that *mind supervenes BNS* is not admitted in the literal sense.

Now let us recall the tenets of the opposite philosophies of mind – idealism and mysticism. They traditionally envisage that mind is a metaphysical being called soul, which interacts with physical body in order to produce the stream of mental states. In ancient period, Plato originally articulated the idealist view about human mind. Descartes followed idealism to advocate the position of body-mind dualism. It is highly controversial because the nonphysical aspect of mind -- i.e. self-consciousness -- is attributed to an inner person called homunculus, which raises many inconsistencies. It may be reiterated that the core issue of idealist interpretation of mind is the metaphysical realism, which holds that mind and body (matter), exist really as independent beings. From these considerations we can conclude that the metaphysical view about mind does not help us in our enquiry.

The key points of the above discussion are presented as below:

- The materialist doctrines of mind called epiphenomenalism and CMF are not competent to explain the nonphysical aspects of mind, specifically the consciousness.

- On the other hand, the metaphysical notion of soul is unsatisfactory for explaining the existence of mind; the realism will lead to the intractable problem of body-mind dualism.
- Since we admit the nonphysical aspects of mind, the issue of body-mind dualism must be solved without adopting realism. It suggests that the dual aspects of body and mind are phenomenal; they are based on the method of our mind for knowing the things of world.

It is necessary to recollect the distinction between body and mind for the proper setting of our deliberation. We can say that the word *body* primarily denotes the inanimate substances like stone, table and chair when these are treated as material (physical) objects. That is, body is supposed to be made exclusively of matter. Additionally, scientists treat the biological body – material structure plus life – of a living organism also as body. Here, body is made of cells where each cell is regarded as a higher organization of certain material molecules. Then life is considered as the byproduct of the material process happening in the cells. But the scientific view about biological body faces serious trouble on account of the reasons presented in the previous chapter. This issue worsens when we consider the phenomenon of mind observable in the case of higher organisms, most significantly in human beings.

Various mental states like emotions and ideas have the wonderful property of *consciousness*, which cannot be reduced to the level of material process. This implies that mind, with the property of consciousness, is a separate entity categorically different from the body. If we treat body and mind as independent entities, it involves the problem of body-mind dualism. Are we justified in holding that matter and mind are separate substances? *The issue of consciousness can be settled only through the solution of body-mind dualism.* This is the objective of the forthcoming parts of this chapter.

In the course of our discussion, we will introduce the seminal idea that a human being has three main levels of existence - *inanimate level, biological level* and m*ental level*. It is envisaged that these levels are systems of matter1-consciousness1, matter2-consciousness2 and matter3-consciousness3 respectively. So we will develop the core areas of *System Philosophy of Mind,* which seek to answer the age-old questions regarding the structure of human mind.

4.3.1 Structure of Human Mind: The System Model [*]

Applying content view for understanding the evolutionary history of nature, it is reasonable to hold that every human being consists of three main levels as outlined below.

The level of inanimate body : The human body is made of cells in various shapes and sizes - there are about 10^{14} (1 and fourteen zeros) cells in our body. Each cell has a concrete structure as well as the property of life. Here we consider inanimate body exclusively as the assembly of the concrete structures of cells, without taking life into account. Note that inanimate body is like dead body and it is similar to a building made of bricks. In an ordinary way we say that the inanimate body is physical body; but there is an essential difference in the meanings of *inanimate* and *physical,* as will be explained in due course.

The level of life: It pertains to the functions of biological body, which is composed of cells and its higher combinations like tissues and organs. We can divide our biological body into two parts called mechanical organ and mental organ. Since *brain and other parts of nervous system (BNS)* is the production centre of mental phenomena, we can treat BNS as mental organ. The other parts of human body such as bones, muscles, digestive system and blood vessels as a whole is called mechanical organ for convenience.

The level of mind: We can show that mind consists of physical and nonphysical parts, which are *the neural networks in BNS* and *the diverse forms of mental states* respectively. The neuron cells of BNS – the mental organ – are the basic units of the physical aspect of neural networks.

But the mental states, produced in neural networks, are nonphysical for reasons explained earlier. In order to deliberate upon the existence of human mind, we must also take into account the other two levels of human being mentioned above.

Now let us consider the second level, i.e. biological body. It is the practice of modern biology to treat both mechanical organ and mental organ as physical structures based on the theory that the constituent cells are made of various macromolecules like DNA and protein. However, we have established in second chapter the mistakes of this scientific view, which ignores the nonphysical aspects in the function of cell. It is the core principle of genetics that all activities in cells are determined by the genetic code (sequence of information) embedded in the structure of DNA. But the creativity of genetic code cannot be explained through physical approach.

We normally believe that mind exists as a separate structure different from body. There are of two main reasons for this belief: Firstly, mind can causally affect the organs of biological body including brain. For example, I get headache due to unpleasant feelings like anger and depression. Secondly, mind possesses the attributes of purpose, creativity and freedom. The steps followed in planning, calculation and decision making involve goal-oriented organization of mental events. Moreover, the aspects of self, personality and moral conscience also indicate that mind exist as a being or entity over and above the structure of neuron networks. In this situation, we have to admit the two-way causal relation between mind and the structure of BNS. ***It was stated earlier that the phenomenon of mind supervenes the neural networks of BNS; but this principle is inadequate. This is the crux of the problem we are addressing now.***

We may focus on the *existence* of mind, which produces the various mental states like emotions, desires, ideas and motivations. As per the above scientific view, the existence of mind depends on the components of biological body such as DNA, genetic code, neural networks and so on. **Extending the critical analysis of Immanuel Kant to our discussion about mind,** we can note that the terms representing such components are predicates, which are defined as the aggregates of particular sets of empirical properties. At the same time such predicates are theoretical entities conceived on the basis of phenomenal observations and logical thinking. In any case, the predicates are deductive concepts pertaining to our phenomenal knowledge. Here we can make use of Kant's assertion that *existence is not a predicate*, it implies that existence is not a property belonging to the set of properties included in the predicate. According to this line of thinking, there is no necessity that a predicate represents a really existing thing. In other words, the theoretical entities used for conceiving the constituents of human things cannot be said to exist really. In the light of the above doctrine regarding phenomenal predicates or theoretical entities, we have to address the question of phenomenal *existence* of mind.

To tide over this problem the best way is to resort to the **system model of organism** introduced earlier in section 2.4. Accordingly, the hierarchical structure of biological body -- including cells, tissues and organs -- is formed by the dialectical relation between physical and nonphysical aspects. *These opposite forces are relative and complementary.* Accordingly, we treat the macromolecules like DNA as physical aspect (X axis); the information or genetic code is the nonphysical aspect (Y axis). So we adopt the production function of macromolecule-information system, in order to show the phenomenal existence of the parts of biological body - all these parts are systems. In the present context, we can rename macromolecule as matter2 and information as consciousness2. This X-Y system is the fundamental cause for the existence of various organs of a human being.

Though it is a bit difficult to understand the **existence of mind as a system,** a simple explanation is attempted here. As an organ of biological body, the BNS is activated through electric impulses associated with sense organs such as eye, ear, nose, tongue and skin. Thus neural impressions are formed in a *disorderly way* in BNS. What is emphasized here is that a higher level of consciousness is interacting with the unorganized impulses of BNS in order to make the desired networks and organize them further in a purposive manner. The functions of sensation, perception and thinking (feeling) occur due to the formation of *orderly neural networks (ONN)*, which serve as the basis for various mental states.

Consider the example of the formation of a company with hundred members. Originally these people are unorganized. But from them, ten smart people come together as promoters to form the company, including the remaining ninety also as members. Then the new company is an orderly and hierarchical organization of hundred members. This instance illustrates the emergence of mind, which is a super structure existing above the unorganized impulses of BNS. The higher level of consciousness is similar to the group of ten promoters of company.

Now we can apply the principle of *system* here. The original unorganised impressions are denoted as matter3. But the formation of specific orderly patterns of neural networks in BNS is achieved by the joint operation of matter3 and consciousness3. Extending the method of system to this higher level, it is proposed that **the level of mind exists as a system of matter3-consciousness3**. Here, matter3 represents the random neural impulses of BNS while consciousness3 represents its complementary part. The advantage of system model of mind lies in the articulation of existence from static perspective.

As the next stage, the new theory modifies our common view that mind exists statically as a being with two levels denoted by unconscious mind (UCM) and conscious mind (CM). Now we have to add *orderly neural networks(ONN)* also as an essential part of mind, which exists as a matter3-consciousness3 system. This content view about mind is similar to that of a factory which produces three kinds of output using separate technologies. As per the input-factory-output scheme, we can note that the mind produces three main classes of outputs namely *orderly neural networks*, *unconscious mind* and *conscious mind*. Let us consider the forgoing fact in a static manner in tune with the content view. Then the unifying illustration of inputs and outputs of mind is named as the **system model of human mind**; it is given in Diagram 1.

Now we may take into account the distinction between real form and phenomenal form of mind. The X-Y coordinate system illustrates the existence of mind in real form; whereas the products - ONN, UCM and CM – represent its phenomenal form. The levels of ONN, UCM and CM together form a system; hence, these components do not exist as separate and independent entities. However, we use separate lines to represent them in the above diagram in a practical manner on the basis of Aristotle's rules of thought.

This system model gives a radically new interpretation to the ordinary view that mind exists over and above the brain. In the course of evolutionary history, the level of mind emerged as a higher system supervening the biological system of BNS. *Remember that by now we have removed the physical version of supervenience doctrine. So we simply say that mind is a company or factory formed by the dialectical relation between the opposite forces of matter (unorganized neural impulses in BNS) and consciousness.*

Chapter 4 / Diagram 1
System Model of Human Mind
(Levels of Human Mind)

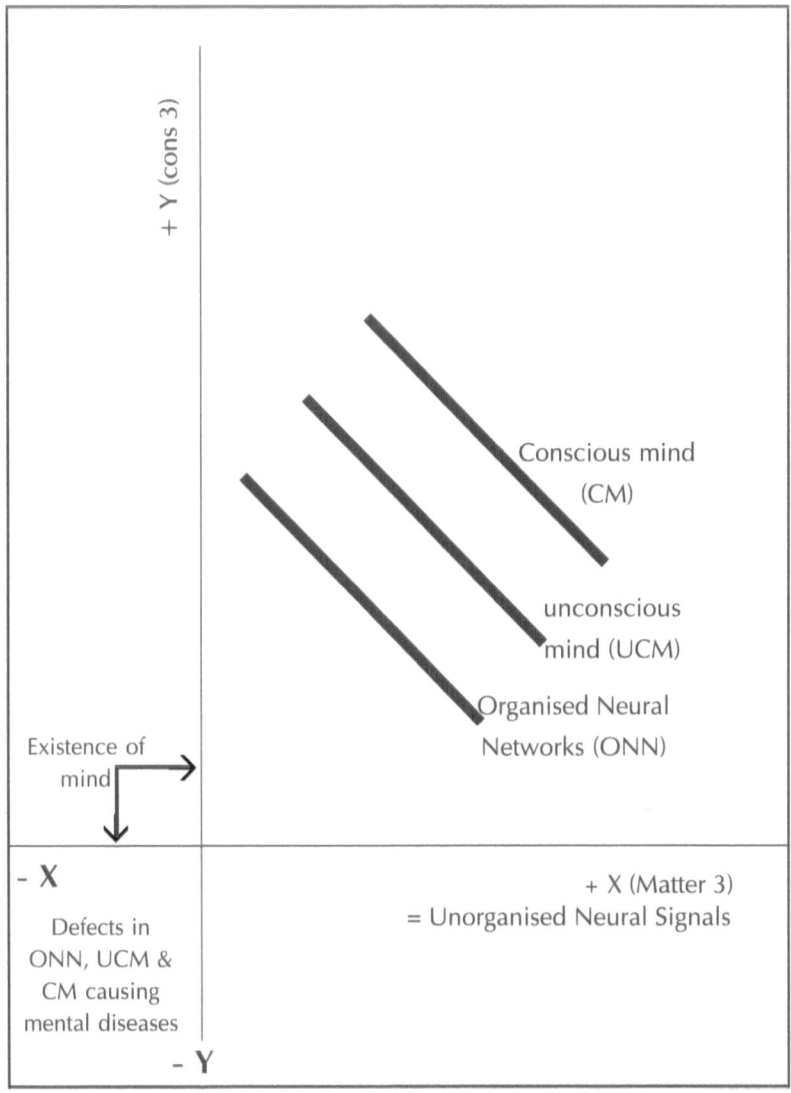

Now we can explain the structure of human mind introduced in first section with regard to the function of generating various kinds of knowledge. As per Table 1 given there, our **cognitive mind** consists of four parts, namely intellectual unconscious mind (A), mystic unconscious mind (B), intellectual conscious mind (C) and mystic conscious mind (D). Here conscious mind and unconscious mind are further divided into intellectual and mystic faculties. But, we need a static model for addressing the issue of existence.

The traditional worldviews and streams of philosophy have failed in connecting the various parts of mind in a comprehensive manner because such philosophies adhere to partial views about the aspects of mind. In contrast, the system model of mind illustrates the existence of levels formed by the productive relation between matter and consciousness in successive levels.

Chapter 4 / Diagram 2
Structure of Human Mind

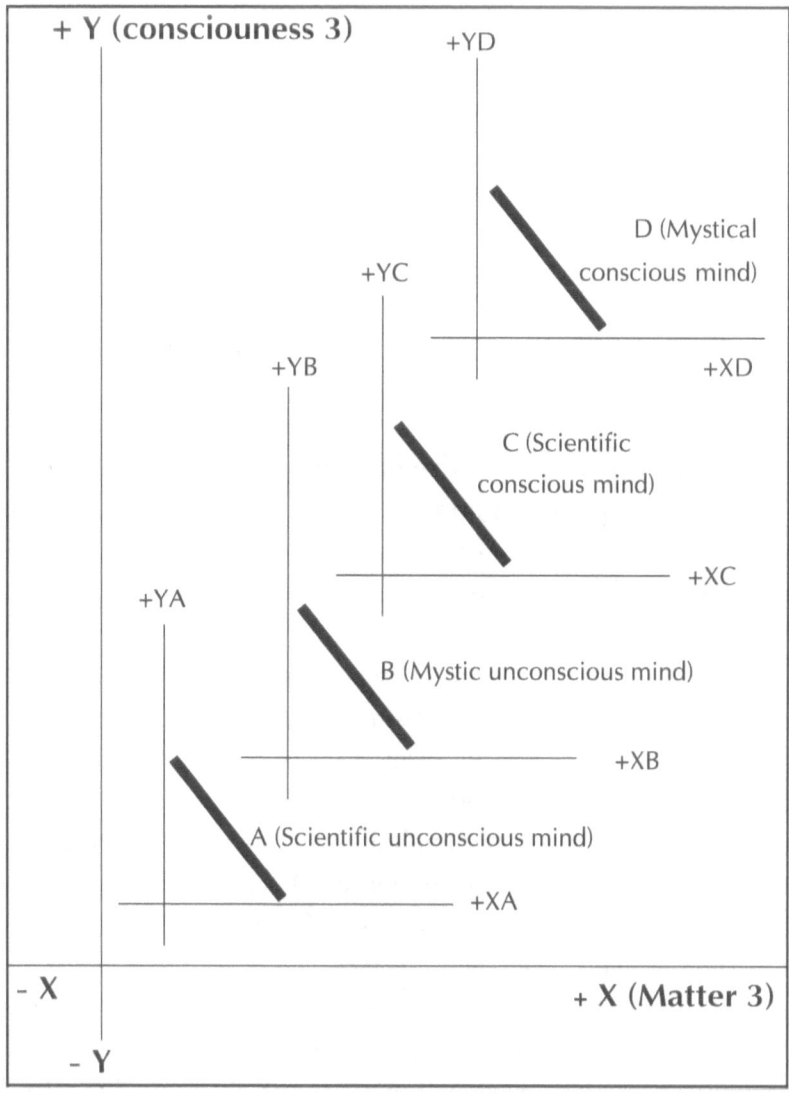

This Diagram illustrates **the system model of the structure of cognitive mind.** Each of the four levels of cognitive mind has the underlying level that is a corresponding segment of orderly neural networks. But *as a convention*, we do not show the divisions of orderly neural networks when we describe the structure of cognitive mind. Hence the segments of orderly neural networks are not explicitly shown in the diagram.

4.3.2 Solution of Mind-Body Dualism

The issue of mind-body dualism assumes central position in modern philosophy as it causes heated debates about the basic components of nature and the way of knowing diverse phenomena. According to the Cartesian view, mind is the same as soul itself with the attribute of rational thought. It is *nonphysical* having the property of self-consciousness. On the other hand, body consists of biological body (physical body + life) together with the sensory experiences. It is a practical convention to hold that biological body is made of matter so as to treat it as *physical*. Additionally, Descartes assumed that animals and lower beings have only biological body, since they do not have rational mind. The most important consequence of *mind-body dualism* is that it generates serious dilemmas in understanding human existence and behavior. If mind and body are separates substances, how can they interact for producing our actions and other bodily movements? In addition to the interaction problem, a lot of other serious issues also arise from the Cartesian theory, as mentioned in the second section.

Through meticulous analysis, we can note the conceptual fallacies inherent in the above notion of Descartes. He considers the mental states in first-person-perspective (FPP), since it is a sequel to his method of doubt. He is absolutely certain in FPP that he has ideas and feelings. But then Descartes shifts to the third-person-perspective (TPP) to hold that he has mind as a thinking substance, as per his axiom phrased as *cogito ergo sum* (I think, therefore I exist). **The mistake committed by Descartes is that he converted his FPP knowledge into a proposition of fact in TPP, applying Aristotle's rules of thought**. And there is no justification to the Cartesian approach.

Taking into account the latest developments in biology and psychology, we can state that the problem of mind-body dualism persists even today when we apply Aristotle's rules of thought to mental and bodily phenomena in TPP. Thus we commonly say that mind exists as a nonphysical being above the physical process of neural networks in brain. It can be emphasized that the distinction between nonphysical and physical things is merely our mental construction; it is not real. We need a new philosophy of mind, in which the realism of mind and body is avoided.

At the outset, we reiterate that the scientific method of viewing inanimate body is to treat it as physical body. As explained in earlier occasion, the term *physical* refers to *that made of matter including energy* which can be quantitatively measured by mathematical means. Examples of important physical concepts are length, matter, energy, volume, weight, motion, electricity, space and time. We need such concepts for studying the physical phenomena. Since matter and energy are inter-convertible, we can simply say that the notion of matter includes energy also. Our subconscious mind has a particular faculty of translating inanimate thing into physical thing; so in our ordinary usage the terms inanimate and physical are treated as synonymous. But the actual meaning of *inanimate* is "that which does not have life". So the distinction between *inanimate* and *physical* must be borne in mind for proceeding with the present discussion.

Using the scientific approach of knowledge, we can treat the first level of human being as physical body. It is a structure made of atoms and molecules which follow the laws of physics, chemistry and related sciences. The various kinds of cells are formed through the organization of water and other molecules like DNA, RNA and proteins. For explaining the formation of various cells we must take into account the phenomenon of life, which has nonphysical aspects also. If we treat the cell as a structure made of atoms and molecules, it is a physical reduction because we exclude the nonphysical aspect of life in cells. In short, physical body is like a dead body observed in scientific way.

When we analyze the structure of atom, there are many serious controversies; detailed articulation of the existence of matter has been given earlier. We have seen therein that the elementary components of atom must be unified by the tenets of *System Philosophy*. As per this thought, for explaining the existence of various atoms, we proceeded philosophically to synthesize the opposites called matter and energy. Then it was proposed that the physical atoms and molecules are formed by the production function of matter-energy system, according to X-Y coordinate model.

However, the scientific description of physical body as matter-energy system does not take into account the aspects of *purpose*, including creativity and freedom, inherent in the constitution of various forms of matter. We may recollect that science deals with the *how* question only, leaving aside the *why* question about phenomena. The notion of physical body excludes the aspect of purpose; hence we must revert to the original idea of inanimate body. Since purpose is nonphysical, it is the characteristic property of *consciousness* inherent in nature.

In this situation we can realize that the inanimate body exists as a system of matter1 and consciousness1; here matter1 is a generic term denoting the property of extension and inertness, while consciousness1 is the source of activity displaying purpose and other nonphysical attributes. For better clarity, the X coordinate of inanimate system is denoted by matter1 while the Y coordinate is denoted by consciousness1. This implies the complementary as well as interdependent existence of matter1 and consciousness1. So it is to be understood that the components of inanimate body are formed by the productive relation between physical and non physical forces.

At the same time, in physical world, energy is responsible for all kinds of activity of matter. So we get the philosophical principle that *matter-energy system is a physical reduction of matter1-consciousness1 system*. Our scientific mind possesses the power of **physicalisation** for reducing the matter1-consciousness1 system into matter-energy system. It generates the notion of physical body made of atoms and molecules. [*].

As per the foregoing, a human being has three levels of systems as marked below:

- *Inanimate level* as the system of matter1-consciousness1.
- *Biological level* as the system of matter2-consciousness2.

- *Mental level* as the system of matter3-consciousness3.

It is possible to state that matter1, matter2 and matter3 are successive organizations of matter. Interestingly, consciousness has ascending order of three levels which interact with the corresponding levels of matter. **In this way, we can hold that a human being exists as a system of matter and consciousness.**

The three levels of human being – inanimate body, biological body and mind – form a hierarchy of matter-consciousness systems. But these three levels of systems are interconnected in a highly mysterious manner. The digestion of food happens in the first level; mainly it is the field of biochemistry. The nutrients are absorbed through blood and used for building cells as well as for giving energy to our organs. The aspects of good health are primarily in the area of biological body but it has implications to the function of our mind. Only a healthy person can think and engage in intellectual activities. In the reverse way, mental disorders like depression and sadness affect the well being of body; it will cause digestion problem, as just one among many kinds of diseases. This philosophical theory is illustrated by the **system model of human being** as shown in Diagram 3 below.

Chapter 4 / Diagram 3
System Model of Human Being

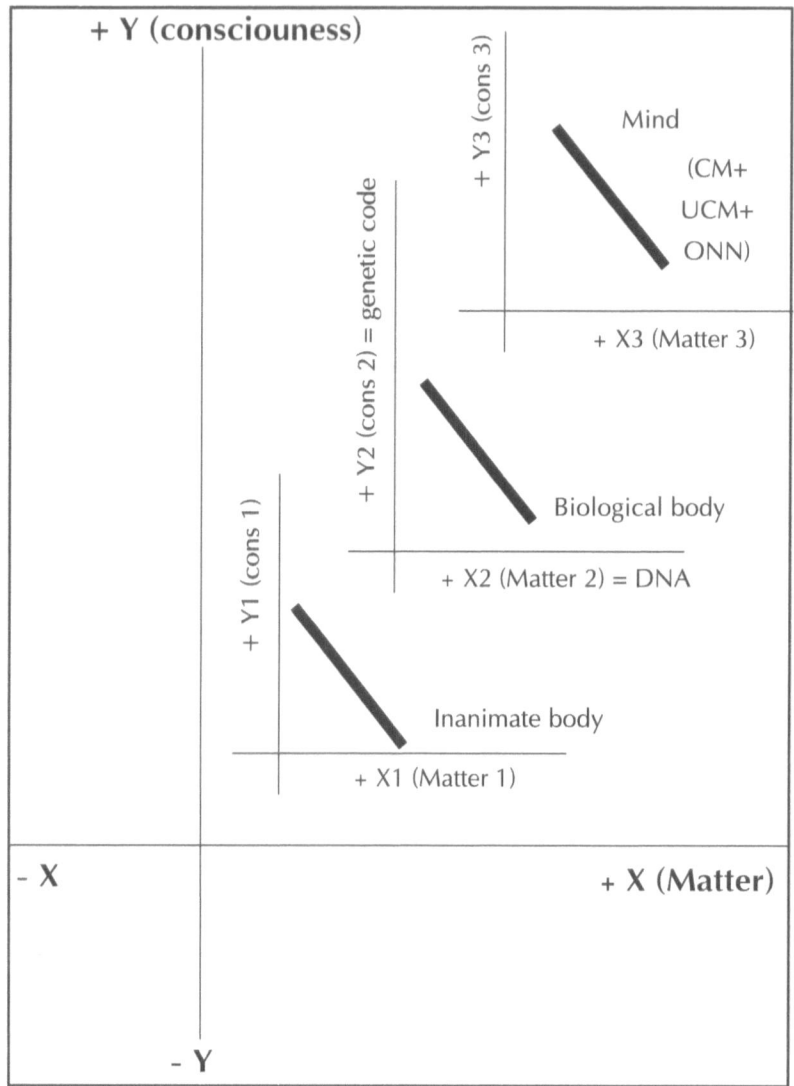

The development of human being from the stage of an embryo is now construed as the dialectical process of the matter-consciousness system. The successive stages of growth, maturity and aging are illustrated by the corresponding versions of the above diagram. We may remember that the system model of human being, involving three levels of subsystems, is a philosophical theory; it contains predicates or theoretical entities. This theory is supported by the stages of hypothesis, deduction and testing. Using the evidences about the interconnectedness of material and nonphysical aspects of mental states, we make the *inference* that mind exists as a subsystem of human system and that it is the productive union of opposite forces called matter3 and consciousness3. *Since existence is an inductive inference, the system model does not involve realism.*

The foregoing system philosophy of mind may be heralded as the solution to the problem of mind-body dualism by virtue of the system models illustrated by the diagrams 1, 2 and 3. Human being is a system of matter and consciousness; this system has three levels of organizations, namely inanimate body, biological body and mind. Accordingly, biological body and mind are separate, but mutually dependent, subsystems of human being. When these subsystems are treated as two levels of matter-consciousness, there is no dualism per se.

However, it is our common sense view to perceive body and mind separately. This is due to the fact that the perception of things is through the application of the rules of thought systematized by Aristotle. Nevertheless, though we observe two things separately, they are interconnected as two subsystems or levels of universe. The problem of dualism arises when we treat the two things as separate and independent in the absolute sense.

We have developed the principles of System Philosophy in order to explain the traditional dichotomy in a new perspective. By virtue of the integrative view of human being, the mind- body dualism is replaced by the matter-consciousness system. **Since a person is a whole with interdependent levels or layers, he has an intuitive conception of the wholeness and it is expressed by the words like *I and self.*** In other words, I and self are theoretical concepts which are used to frame sentences in FPP. Based on these grounds we can firmly say that the phrases like *I think, my body* and *my mind* are mere practical ideas formed in our mind, without any connotation about mind-body dualism.

4.3.3 What is Consciousness?

A sound theory about consciousness must adopt the following tenets:

- The FPP aspect of consciousness is to be considered in a **secular** manner without the influence of religious assumptions. At the same time, we want to avoid realism and synthesize the nonphysical and physical aspects of mental state.
- We have to define consciousness in a broad manner so as to encompass the spectrum of living beings. It must be admitted that attributing *self-consciousness* only to human beings is a linguistic practice we adopt.
- More over there is a need to integrate unconscious mind and conscious mind so as to agree with the findings of cognitive psychology. Hence, the division between unconscious mind and conscious mind may be renamed as the division between *low conscious mind (LCM) and high conscious mind (HCM)* respectively. However, as a custom, we will continue to call the HCM of human beings as self-consciousness.
- We must adopt content view for articulating the existence of mind with the hierarchy of levels of consciousness.

These stipulations take us forward in the path of *system philosophy of mind*. The emerging vision about consciousness is presented here in a brief manner culling out from the details given in the previous sections.

We may recall the Diagram 1 showing the *System Model of Human Mind*. Renaming UCM and CM as LCM and HCM respectively, we can integrate the two levels of mind. So the drastic separation between UCM and CM is avoided. In other words, we can divide mind into two subsystems – LCM and HCM. This shows that two levels of consciousness3 together with the corresponding levels of matter3 would produce the LCM and the HCM respectively.

Mind, which is the totality of LCM and HCM, exists over and above ONN. But always we have to consider LCM or HCM in combination with particular segments of ONN. That is, each of the levels LCM and HCM has two complementary parts – a segment of ONN and a corresponding level of consciousness. *There is no justification in holding that BNS is a physical object and it is separate from nonphysical consciousness.*

Further, we can give similar interpretation of the production of various mental states adopting the process view about mind. The system model is like an input-output system. The factory called mind produces various mental states as outputs so that they have the dual components of matter and consciousness. The various mental states can be represented by sublevels of *low conscious mind* and *high conscious mind*. The above model is applicable to the case of nonhuman organisms also, which have BNS. They too have two levels of mind, LCM and HCM, each of which can be illustrated as X-Y systems. But here the HCM does not have self-consciousness as per our norm.

As conclusion, the essence of the foregoing exposition may be listed as follows:

> ➤ The central objective of the **system philosophy of mind** is to examine the doctrine of neuroscientists that mind exists above the brain and other parts of nervous system (BNS). The scientific view does not recognize the nonphysical aspects of mind like creativity, purpose and freedom. It leads to the conflict with the religious theory that mind exists as a metaphysical being; this is traditionally called as the mind-body dualism.
> ➤ By virtue of the system models (Diagrams 1, 2 and 3), adopting *content view*, human being is a system of matter and consciousness; this system has three levels of organizations, namely inanimate body, biological body and mind.
> ➤ Accordingly, inanimate body is a matter1-consciousness1 system and biological body exists as matter2-consciousness2 system, whereas mind is a matter3-consciousness3 system. Here matter3 is the set of unorganized neural signals in BNS. The implication is that biological body and mind are separate, but mutually dependent, subsystems of human being. When these

subsystems are treated as two levels of matter-consciousness, there is no dualism per se. Hence the age-old problem of mind-body dualism is solved.

➢ As per process view, we can treat the mind as a factory with X-Y model, which produces three main classes of outputs namely *orderly neural networks, unconscious mind* and *conscious mind* – these are interconnected.

➢ Each mental state is the combination of the said three main classes of outputs. Now it is necessary to take an integrative view of consciousness. Hence, we propose that the division between unconscious mind and conscious mind may be renamed as the division between *low conscious mind (LCM) and high conscious mind (HCM)* respectively. It is our convention, in the case of human beings, to say that the *high conscious mind* possesses self-consciousness involving the use of language and grammar for expressing ideas. So we naively hold that the lower animals and plants do not have self-consciousness and that they possess *low conscious mind* only.

NOTES of Chapter 4.

1   ***The special references used for discussing the issues under philosophy of mind are:*** Brennan (2005); Feser (2006); Grayling (Editor) (1995); Guttman (2007); Heil (2003); Leahey (2005); Maslin (2001); Solso (2005)

2   Detailed description of Cartesian body-mind dualism is available in Heil (2003), Feser (2006), Leahey (2005) and Maslin (2001). George Thomas White Patrick (1978) mentions in *Introduction to Philosophy, page 293,* that the mind-body problem is one of the "seven world-riddles", which have been said to be incapable of solution.

\# 3   The structure of human *brain and other parts of nervous system* (BNS) is discussed in detail in Ammar Al Chalabi *et al* (2007), Guttman (2007) and Solso (2005).

\# 4   Here we focus on the functions of unconscious mind and conscious mind theoretically, rather than describing the corresponding parts of brain.

\# 5   Main references for the various aspects of consciousness are Feser (2006) and Maslin (2001). The Table 1 showing the function of cognitive mind is in accordance with the scientific theory that mind exists as a level above the neural networks in brain. This diagram is an original idea.

\# 6   The structure of cognitive mind, as shown in Table 1, can solve the mystery about the nature of our acts under religion and art. This frame work of linking intellectual mind with mystic mind is my original idea.

\# 7   The table showing the worldviews is an important original idea paving the way for System Philosophy.

\# 8   The Indian philosophy of Vedanta also describes the chariot model of mind. Explanation of the notion of *homunculus* is available in Leahey (2005) and Maslin (2001).

\# 9   We must note that the Aristotle's rules of thought constitute the reason behind the notion of mind-body dualism. It is my original idea to link the rules of thought to Cartesian dualism and solve the problem of dualism in this chapter.

\# 10   Macquarrie (1985) gives a good account of pantheism. Heil (2003) and Maslin (2001) have described parallelism systematically. The definition of panpsychism is obtained from Griffin (2000). Refer Leahey (2005), page 356, for the modern version of panpsychism, which is developed by William James

(1842-1910) through physical process view – it is popularly called as functionalism.

11   Chardin (1965), *The Phenomenon of Man*, (Harper Torchbook Edition, New York, 1965)

12   Romijn, Herms (2002), *Are Virtual Photons the Elementary Carriers of Consciousness?*

Other important thinkers articulating quantum consciousness are Ken Wilber, Wilder Penfield, Rupert Sheldrake, Nobert Wiener, Humberto Maturana, Francisco Varela, Gregory Bateson, George Wald and Ervin Laszlo.

13   Grayling A.C. (Editor) (1995), *Philosophy: a Guide through the Subject*, page 254.

14   Brennan (2005) and Leahey (2005) provide the details of the contributions of the pioneers in the field of functionalism, psychoanalysis and later theories under modern psychology. The various philosophical doctrines of mind are compared thoroughly in Heil (2003). However, the definition of physical process view and the criticism of concerned doctrines is my original idea.

15   See Feser (2006) and Solso (2005). The notion of *language of thought* refers to the capacity of producing ideas or feelings without language; it is the function of the unconscious mind working as central processing unit. In the book of Solso, language and cognition are discussed from the point of view of neuroscience.

16   John Searle presented the thought experiment famously known as *Chinese room*; it demonstrated that functionalism cannot capture the qualia and intentionality of mental states. See Maslin (2001) pages 154-160 and Feser (2006) pages 121-125.

The following criticism about the process view of computer model functionalism is my original idea.

17 The description of system philosophy of mind presented in third section – including the Diagrams 1, 2 and 3 - constitute my original idea about the existence of human mind.

Chapter 5

System Philosophy of God and Evil

****

5.1 Anatomy of a Crisis

5.2 Religion is a Social System

5.3 God and Soul are Mystical Concepts

5.4 Justification of Religious Knowledge: Does God Exist?

Author's main original ideas are marked by [*].

The mark [#] gives the number of note at the end.

****

So far our philosophical enquiry was about the scientific knowledge pertaining to physical world, life, evolution and mind. We have found that the fundamental concepts of these topics are not agreeing with the religious belief about natural world. We are perplexed by the consequent conflict between science and religion and desire to rectify it. The discussions of this chapter and the next are meant for that purpose. But it is presented in a concise manner only, because a detailed treatment is reserved for a later book.

In this chapter we are concerned with *philosophy of religion*, which is defined as the epistemology of religious knowledge. On account of divergent worldviews, this subject has failed to suggest a method for unifying the various religions (forms of worship and theologies). These separate religions appear to have different sets of beliefs and activities of worship. *Theology* literally means discourse about God and it includes the rigorous interpretations of beliefs and practices for defending a particular religion or sect. So, we must note that *philosophy of religion* has reached **a state of crisis,** which is analyzed in the following section. At present we have chosen to deal with this subject *in the context of Christian religion*, since it has well developed theological doctrines. The philosophical dilemmas in this field are given below. [# 1][*].

5.1 Anatomy of a Crisis

With reference to the historical development of Christian theology, firstly we have to study the early development of **theism**, which is the theology about transcendent God. We will explain now that theism can be divided into rational and empirical parts. [# 2][*].

We can give the basic classification of the spectrum of theologies under Christianity, in an innovative manner considering the structure of our mystic mind. The four blocks of theologies are constructed using the concepts of rational faith, empirical faith, content view and process view as shown in the following Table.

Table 1 : Religious Knowledge under Mystical Mind [# 3][*].

	Content view	**Process view**
Rational faith (DP)	*Rational Theism* Eg: Roman catholic theology	*Process Theology (including pantheism and panentheism).* Eg:Theology of Protestant Church, Buddhism, Taoism
Empirical faith (IP)	*Empirical Theism* Empirical description of religious worship allied to theism	*Mystic Worship Theology* Special mystic practices like Kabbalah, Sufism, Christian mysticism and Hindu mysticism

In the 20th century, there is considerable diversion from the core concerns of philosophy of religion by taking the route of process view to club religious matters with scientific propositions. [# 4].

Reviewing the literature pertaining to philosophy of religion, we can observe that the subject is facing a crisis now. **The ensuing paragraphs would present the key points about the anatomy of crisis.**

Rational faith has many philosophical issues, since Immanuel Kant trounced the validity of metaphysical realism. Rational ideas about God are not supported by sensory evidence; hence it does not imply the real existence of God. On the other side, the set of empirical propositions, pertaining to religious belief, contain the claims about the existence as well as the attributes of God. Then the moot point is whether the bodily experience of worship can lead to any abstract principle about divinity. It involves the deeper question: Can we get abstract ideas about God from sensory experience? This is the crux of the conflict between rationalism and empiricism, when we consider religious knowledge. It implies that the above table coherently illustrates the main controversies and problems with regard to both methodology and source. We can give the list of principal issues as below:

1. The synthesis of rational faith (DP) and empirical faith (IP) has not been achieved so far.
2. The empirical aspect of religion is largely ignored in the theological studies, since theology has been developed as a form of rational thought.
3. There is an urgent need for a proper philosophy of mind that reconciles the rational and empirical aspects of mystic mind.
4. The dichotomy between content view and process view must be removed. For that purpose, the dual aspects called transcendence and immanence need to be integrated by proposing a new doctrine of God.

There are many cogent arguments, thanks to the critical philosophy of Immanuel Kant, to establish that God or mind does not really exist as a substance. Having rejected the religious version of reality, we may look to the other side where materialism holds that matter is the reality. A group of scientists and other like-minded people unleashed the campaign called **atheism**, for attacking religion describing it as the set of non-rational and meaningless beliefs. Typically, scientists have displayed disrespect for philosophy by insisting that matter is the only reality. They assert the position of **scientism** meaning that scientific method and laws give the exclusive way to attain the truth about universe. The key points about the development of atheism are postponed to next chapter.

The successive chapters of my books aim to advocate the view that neither mind nor matter can be accepted as the ultimate form of reality. That is, none of these single principles is suitable for explaining the natural phenomena. Hence, the conflict between theism and atheism cannot be resolved by favoring one or the other. In this situation the best alternative is to adopt the system model of ultimate reality; this topic is postponed to the next book.

However through ingenious arguments, we can show that the abstract concepts - definitions and attributes of metaphysical beings - of religious knowledge have justification, provided religion has existence. That is, if religion exists as a social system, the propositions like "God exist" and "God is love" have validity within that framework. This principle can be established in the ensuing sections.

5.2 Religion is a Social System

We require a philosophical method to explain the existence of religion with good and bad aspects. If theology is meant for the advancement of devotees in the good direction of spiritual path, why do evils occur in religious communities? To answer this question, we must link fact with value in the case of religious system. Every factual aspect of a religion - such as the worship, prayer, ritual and so on - can be considered as a voluntary action. **That is, religion is an action system**. Moreover, religious actions are performed by the goals of our mystic mind. Every group or society is formed on the basis of a set of established rules or values, which are collectively represented by the term *institution*. When we aggregate the institutions developed by human civilization, there are seven non-overlapping institutions at the global level. Such macro institutions are appropriately called as *Seven Life Systems;* this idea is briefly explained as following.

The unconscious part of human mind is the source of purposes and motivations, which are responsible for the formation of our social systems. We may analyze the resemblances and differences of the multitude of such manmade social systems. In this way we can classify the entire spectrum of social systems into seven macro systems -- natural life system (NLS), economic life system (ELS), political life system (PLS), family life system (FLS), ethical life system (ETLS), religious life system (RLS) and artistic life system (ALS). The system model of human mind shows that these seven life systems are produced by separate faculties of human mind, according to the respective aims of practical life.

System Philosophy gives a radically new vision about the existence of *religious life System* (RLS), which is the totality of all religions as social systems. The religious faculty of human mind has the special purpose of experiencing spirituality in natural things. Accordingly, it gives spiritual roles to specific things so that they can be called as *religious objects*. The main examples of religious objects are churches, priests, holy things like cross and religious books as well as prayers and other forms of worship. The interrelated existence of a group of religious objects constitutes a religion as a social system. Various religions exist *ontologically and scientifically* because they are systems of the religious objects formed by the mystical faculty of human mind. It may be added that religions as well as constituent objects – being systems themselves - have *phenomenal existence.*

Here we can derive the structure of RLS formed by the specific goals of religious part of mystic mind. (Note that mystic mind has religious part and artistic part). Every goal is a value, which has the dual aspect of self-interest (SEI) and society-interest (SOI) running in opposite directions of good and bad. Taking SEI as X axis and SOI as Y axis, the existence of Religious Life System (RLS) is illustrated by the X-Y coordinate model. Various religions are produced by the dialectical relation between SEI and SOI as per the system model of RLS. It establishes convincingly that every religion has good and bad parts, which appear in the first and third quadrants respectively.

Finally we can note that the philosophers of religion have struggled without success to propose a philosophical frame work for *religious pluralism*. Now it is an enlightening principle that RLS grows like a tree with recurrence of ever new branches and leafs. The diversity of theologies and modes of worship are determined by historical and social factors, which are represented by the concepts of SEI and SOI. [# 5].

5.3 God and Soul are Mystical Concepts

We are now concerned with the mystical facts under content view produced by believers participating in RLS – such facts pertain to transcendent God. **The topic of existence can be discussed only through content view.** For example, consider the propositions: *God exists, soul is immortal, God is love* and *I escaped from the accident due to the mercy of Mother Mary*. These pieces of knowledge are inferences produced by connecting the cognitive processes of deduction and induction. In this context it is necessary to recall the theory of knowledge pertaining to science. We will now pursue the epistemological analysis of religious knowledge in a similar vein.

We may state that it is a historical irony to use the word 'theism' to the rational theism exclusively. Our analysis shows that the radical separation between rational faith and empirical faith – that is between DP and IP – had caused lot of confusion in the erstwhile history of philosophy of religion. In order to understand this point, let us contrast Plato's theory of mind with the concerned doctrine of System Philosophy.

We consider here Plato's famous theory of Divided Line. Plato clubbed intellectual mind and mystic mind, by placing intuition above intellect while imagination (faith) is kept below perception; thus he accounted for the four levels of Divided Line in tune with the teleological view of idealism. Plato failed to recognize that intellectual mind and mystic mind are parallel faculties of mind. This brought about far reaching consequences in the further course of Western Philosophy. Since he treated God, or Idea of Good, as a rational principle determining the classes of phenomenal world, the notion of reality was transformed from secular to religious realm. This became the theory of reality under *organic world view* or idealism, which is in constant conflict with materialism. In modern period, it paved the way for serious dispute between religion and science with regard to the ultimate questions about universe.

In the light of System Philosophy of Mind, presented in previous chapter, we know that our cognitive mind can be divided into intellectual mind and mystic mind, where each has two levels namely rational level and empirical level. Then the four parts of Divided Line can be rearranged into a 2x2 table as blow.

Table 2 : System Philosophy of Mind and Plato's view [# 6][*]

	System Philosophy of Mind		PLATO
	Intellectual mind	Mystic mind	
Rational part	Intellect	Intuition	Intuition
			Intellect
Empirical part	Perception	Imagination	Perception
			Imagination

As per the foregoing, the practice of hitherto philosophy of religion was to treat rational faith (DP) and empirical faith (IP) as separate and contrasting descriptions of spirituality. Consequently, theism evolved along parallel paths; rational theism neglected empirical theism and vice versa. The philosophical issues of this dichotomy still remain unsolved. However, the fundamental tenets of *system philosophy of mind* would enable us for an integrative view about methodology and source of theism.

It is an important insight of *System Philosophy about knowledge* that every meaningful proposition exists as a TyHDTI scheme, which can be decomposed into DP and IP. Accordingly, we assert that DP or IP taken in isolation does not qualify to become knowledge, because *there is a holism about meaning* contributed by deduction and induction. Meaning is a whole; hence a deductive or inductive part is understandable only in its relationship to whole. [# 7].

For example, under content view, the proposition *God exists* is the merger of the two sets of ideas as below:

- The abstract ideas deduced rationally about the definition of God and divine attributes. This set consists of Ty, H and D. Theology constitutes the *theory part* of religious knowledge under TyHDTI scheme.
- The empirical or inductive ideas pertaining to religious experience about the existence of God. The propositions under T and I are included here.

So we reach at the solution for the outstanding issue of methodology and source; that is, the reconciliation of the conflict between rationalism and empiricism. The mode of thought pertaining to science and philosophy may be described as *intellectual mind*, in order to separate it from the *mystic mind* applied in art and religion. When we advocate that our mental faculty exists as a system of rational part and empirical part, every inference or knowledge has an integrative structure. The **system model of religious knowledge** can be constructed now. It establishes the complementary existence of rational theism and empirical theism for generating meaningful propositions about religion under content view.

The integration of *process theology* and *mystic worship theology* is also possible along the above lines, but it is the process version of God and related aspects. The knowledge about immanent God is the system of DP and IP levels as per TyHDTI scheme. Hence, we can note that the *system model of religious knowledge* reconciles the issue of **dialectical theism**, which denotes the dual aspects of transcendence and immanence of God. [# 8].

The method of System Philosophy is to synthesis the rational part and empirical part of theological doctrines. Then we have to understand God in two complementary ways called content view and process view, which can be further synthesized as per the following section.

5.4 Justification of Religious Knowledge: Does God Exist?

In the next step, we intend to find the justification for the metaphorical inference about God's existence. *Since justification is a part of secular epistemology, it has scientific connotation.* We are conducting an objective and disinterested enquiry whether religious beliefs can be justified. It is now convincingly clear that the existence of Religious Life System (RLS) is the justification for the mystical propositions about the definition of God and other aspects of religious worship. This point is elaborated in the scheme given below.

❖ The mystical faculty of human mind has the dual purposes of self-interest (SEI) and society-interest (SOI), which together serve like a manufacturing process to produce RLS. Hence RLS exists scientifically in the form of X-Y coordinate system. As an alternative interpretation, we can say that the empirical and rational aspects of mystic experience of believers work dielectrically to produce RLS, which is the *religious reality*. We may add that RLS has positive and negative parts, representing respectively the goodness and evil associated with religious activities.

❖ When a believer participates in the rituals and traditions of RLS, he or she gets mystical experiences or ideas subjectively. For generating religious propositions under FPP and subsequently under TPP, the worshipper makes use of theory or teachings of theologians. Through the successive stages of hypothesis and deduction, the devotee arrives at the rational propositions like God exists, Evil exists, God is love, heaven and hell exist, God saved me from the accident, and so on.

❖ Through repeated activities of worship, the devotee gets experimental evidence for the said rational propositions. This is the stage of testing for the verification of beliefs. If the strength of evidence is not sufficient, the person may reject the concerned belief.

- ❖ The confirmed beliefs can lead to suitable inductive inferences. **"God exists" is *the most basic inference*** arrived at by the devotee through the participation in RLS. Simply, when the deductive propositions are verified through activities of worship, mystical knowledge is produced - this cognitive process is denoted by TyHDTI scheme.

We have explained earlier that various religions and sects have been evolved in historical and social context. The totality of all religions is the RLS, which is like a tree. The mystical minds of human beings in different communities combined the notions about supernatural forces with a host of myths and stories in order to form separate definitions of God, Evil, soul, heaven, hell, and so on. The structure of worldviews proposed in System Philosophy is the appropriate method to classify the spectrum of religions and theologies.

Through epistemological deliberation, we explained that religious knowledge is ultimately aggregated to two branches namely content view and process view, which are in accordance with the dual aspects - transcendence and immanence - of God. The scientific existence of RLS is the justification for mystic ideas with respect to God and other related entities. Going backwards, the RLS and other six life systems exist by the structure of corresponding faculties of human mind. The phenomenal existence of Nature – including inanimate and living things – is the product of the Ultimate Reality or *paramporul*, which is an X-Y coordinate system of body and consciousness.

Accordingly, the system model of Ultimate Reality explains the production of all inanimate and living phenomena, which can be classified into two classes of good and bad systems as per our ethical perspective. Good and bad has ontological existence manifested in phenomenal systems of Nature.

In conclusion of this analysis, the existence of God is an inference confirmed by the mystical experiences of a worshipper. Holding that the existence of God is not part of the theory (rational theism), the problem of metaphysical realism is removed. As a form of mystical inference, the notion of divine existence is assigned metaphorical and subjective meaning. But the existence of RLS is a scientific fact, which accounts for the production of mystic experience in the worshipper.

Our religious experiences invariably contain the inferences about the dual concepts of God and Evil. No religion can teach about God, without invoking the opposite idea of Evil. However, due to the influence of metaphysical realism about God, theologians appear to fumble when they deal with the role of Evil. They take the escape route of saying that Evil is a fallen angel or an illusion. For understanding the religious notions of God and Evil, we must climb the ladder of System Philosophy so as to distinguish between scientific and mystic areas of knowledge.

The mode of thought pertaining to science and philosophy may be described as *intellectual mind*, in order to separate it from the *mystic mind* applied in art and religion. Using intellectual mind, we can argue that the concepts of God and Evil must account for the opposite aspects of good and bad, experienced in the world from mystical perspective. Further, world has the duality of material and mental parts where the latter is normally called as consciousness. Hence, both God and Evil are composed of material and mental aspects.

It is now possible to remove the controversies about the religious notions of God and Evil, using our intellectual mind. So we can adopt the ***System Model of God and Evil,*** which shows God in first quadrant and Evil in third quadrant of X-Y coordinate system.[# 9][*].

This diagram, showing God and Evil as opposite religious realities, solves the traditional ***problem of evil***. We have many instances where good and pious persons are killed in accidents or natural disasters. **In order to explain mystically the numerous types of evils in the world, it is necessary to hold that such events are the actions of Evil. We should accept that God is the source of all good aspects exclusively; God is not responsible for any evil event. This is the**

solution to the *problem of evil* through intellectual logic. Further it is possible to envisage that there is a dialectical relation between God and Evil, as per the above system model. Our good actions will be like voting for God, while we favour Evil through our bad actions. God and Evil do not exist as separate entities; but the *system model* implies complementary existence of these two opposite forces without attracting realism.

We have the experience of bad things in the world – this leads to the knowledge about Evil. If our mystic mind is cultivated or trained to absorb the *System Model of God and Evil*, then the entire structure of theology can be recast. Moreover, a religious person would not have any difficulty in assuming that God (or Evil) has both material and mental aspects. Note that the mythological stories invariably include good and bad characters having both body and mind; it constituted the early form of theology.

The traditional opposition between idealism and materialism is the reason for the convention of treating God, Evil and soul as spiritual / mental entities. System Philosophy provides the intellectual method for reconciling the opposite concepts of body and mind using X-Y coordinate system. However the hitherto developed theology, which assumes that God, Evil and soul are mental entities only, can still hold good because the importance and attributes given to the various supernatural entities are at the discretion of mystic mind. Such religious knowledge is justified by the scientific existence of Religious Life System.

FAQ about the existence of SOUL

The concept of *soul* essentially pertains to *theism* that is the religious philosophy upholding the existence of a perfect, all-powerful and personal God. In this context, soul is an immortal and metaphysical being originating from God. On the other hand, the process theology adopting mysticism considers soul only as an activity, not as a being. In the following paragraphs, *we will deliberate upon the philosophical issues of the theist notion of soul.*

Life and Mind

Scientifically we can observe that all living beings have various measures of self-consciousness. However, it is a religious convention to hold that human mind exclusively has the property of self-consciousness. This religious view of *human soul* as the source of self-consciousness causes serious confusion. Now we may hold that it is the religious interpretation to say that life and mind are manifestations of the interaction of soul with material body. Further it is asserted that human being has only three parts - body, life and mind – from scientific perspective.

In order to avoid the repetition of points already given in previous sections, the key arguments are presented below briefly.

- ➤ The propositions about soul are produced in our mystic mind according to TyHDTI scheme, which exists as a system of deductive ideas and inductive ideas of mystic mind. So we reach at the solution for the outstanding issue of the reconciliation of the conflict between rationalism and empiricism.
- ➤ When a believer participates in the rituals and traditions of Religious Life System (RLS), he or she gets mystical experiences or ideas subjectively. Through repeated activities of worship, the devotee acquires experimental evidence for the rational and mystical propositions about soul. Simply, the deductive proposition that soul exists is verified through activities of worship -- this cognitive process is denoted by TyHDTI scheme.
- ➤ Thus System Philosophy of knowledge asserts that **the proposition 'soul exists' is a mystic inference** under TyHDTI scheme of religious knowledge; this proposition should not be treated as part of theory (theology). It has been convincingly established that the existence of Religious Life System (RLS) is the justification for the mystical propositions about soul. We may emphasize that the justification is achieved through *secular way* even though soul is a religious concept. System Philosophy does not resort to metaphysical realism.

To conclude this treatise, let us mention about the limitation of neuroscience. According to neuroscientists, experimental methods can show the different sets of neuron networks pertaining to various kinds of knowledge produced in human brain. But ideas (including emotions and feelings) are nonphysical without specific spatial location. Even though neuroscientists may show that certain area of brain corresponds to religious faith, they cannot identify the production of concerned specific ideas. In this situation a scientific project to study the existence of religious entity like soul would be futile.

NOTES of Chapter 5

#1. The main points of this chapter can be treated as my original ideas, since it draws heavily from the tenets of System Philosophy developed in the previous parts of this book. However, this author is especially indebted to the following books of reference:

Anthony Harrison– Barbet (1990); Armstrong, Karen (1998); Caputo (2013); Copleston, Frederick S.J. (1994); Davies, Brian (2000); Dawkins (2007); Esposito, et al (2008); Frost. S. E (1989); George Thomas White Patrick (1978); Grayling A.C. (Editor) (1995); Grayling A.C. (Editor) (1998); Griffin (2000); Hick (1994); Kant (2003); Lavin. T. Z. (1989); Macquarrie (1985); Masih .Y. (1995); Max Charlesworth (2006); Tarnas (1991); Thilly (2000); Thomson (1997).

2. The distinction between rational theism and empirical theism is my original idea. It helps us to solve the long standing issues related to theism. See also Esposito, et al (2008).

3. This 2x2 table is in accordance with the classification of worldviews already presented. It is the key tool for unifying the diverse theologies.

4. Here we may mention about *critical realism*, which is proposed as the method for linking process theology with scientific knowledge. Critical Realism will be explained in the next book, *Discovery of Reality*.

5. Religious pluralism is discussed in Hick (1994), *Philosophy of Religion*, page 117. Its philosophical basis is originally presented here through system model of RLS.

6. Plato mentions about Divided Line in Book VI of the *Republic*. See the description in Lavin (1989), pages 31-42.

This Table 2 for comparing Plato's Divided Line with System philosophy of Mind is my original idea.

7. In the context of linguistic analysis of scientific propositions, W.V.O. Quine introduces the idea of "holism about meaning". See Alex Rosenberg (2000), *Philosophy of Science - a Contemporary Introduction*, page 151.

8. For the discussion about dialectical theism, see Macquarrie (1985), page 54.

9. The separation between *intellectual mind* and *mystic mind* is vitally important for understanding this model.

Chapter 6

Science-Religion Synthesis

****

6.1 Atheism is a Delusion

6.2 Synthesis of Science and Religion

Author's main original ideas are marked by [*].

The mark [#] gives the number of note at the end.

****

It is expedient to conclude this book by indicating how System Philosophy can synthesize the obviously different kinds of knowledge under science and religion. The challenging problem is briefly mentioned below.

Science denies the existence of supernatural beings or forces. It reduces the natural things, which are actually composed of body and mind, into forms of matter and energy. That is, science holds that everything in the nature is *physical*. It formulates cause-effect relations, known as *physical laws*, based on sensory or experimental evidences about the properties of physical things through our *scientific mind*. This is public knowledge in third person perspective. Adopting scientific method to enquire about the origin of universe, a group of scientist has proposed **atheism**, which is the position that matter is the reality and God does not exist.

On the other hand, religion presumes that supernatural beings or forces exist and they cause the affairs of natural world. This is based on mystic experiences through revelation, emotions, ecstasy and meditation of religious leaders as well as ordinary believers. The mystic ideas are private and beyond sensory experience - it is the function of *religious mind*. In this way the doctrine about God is advanced; it essentially involves the conflict between science and religion.

6.1 Atheism is a Delusion

Drawing energy from the disputes about the existence and attributes of God, the intellectual movement called atheism gained strength and popularity in the western world. Without considering its historical development, we may give the central argument of atheism as per the following propositions.

- ➤ There is no scientific evidence for the existence of God. Hence God does not exist.
- ➤ Idealism is rejected; materialism is the true theory of reality.
- ➤ Science is based on materialism and it is exclusively the true form of knowledge.

Atheism is the view that there is no God or other supernatural beings. Additionally, it seeks to explain the events of nature only through physical way of scientific methods. Atheists oppose process theology (pantheism) also since they adhere to materialism. In retrospect, David Hume (1711-1766) had expounded empiricism and materialism as the premise of scientific method; God has no role in his scheme of knowledge. Charles Darwin (1809-1882) is considered the pioneer of atheism in modern period since he tried to explain biological evolution as a material process, without resorting to the notion of God or intellectual designer. Subsequently, Karl Marx (1818-1883), Friedrich Nietzsche (1844-1900), Emile Durkheim (1855-1917) and

Sigmund Freud (1856-1939) developed natural interpretation of social phenomena upholding the project of atheism.

A section of atheists would say publicly that their position is to refute the religious belief in personal God (theism). This involves confusion whether they support the idea of immanent God as typically belonging to pantheism. We can clear this issue by noting that atheists vouch for materialism as theory of reality.

In 20th century, the philosophic doctrines like logical positivism, existentialism and linguistic analysis have contributed greatly for increasing the number of atheists. According to current sources, the followers of atheism constitute roughly ten percent of world population. In this situation, the writings of Richard Dawkins, especially his book *The God Delusion* (2007), require special mention. Dawkins resorts to Darwin's physical theory of evolution to hammer the view that a personal God does not exist. However, we have in chapter 3 convincingly argued that the materialist theory of Darwinism is defective from philosophical perspective. The notion of *natural selection* actually contains nonphysical aspects of purpose and creativity. Nevertheless, biologists treat natural selection as a physical process in tune with the approach of science. So we must conclude that the attempt of Dawkins to promote atheism stands on unsound premises.

At the same time, it may be reiterated that the theological doctrine of evolution – evolutionary theology including *intelligent design argument* – also can be refuted through proper analysis.

But the arguments of System Philosophy about the concept of God as presented in the previous chapter, as well as the topics developed in earlier chapters, would serve to expose the fallacies of both the doctrines, atheism and evolutionary theology. Specifically, the system models of organism and biological evolution together show that we can secularly take into account the nonphysical aspects pertaining to the process of life and evolution. [# 1].

6.2 Synthesis of Science and Religion

The ideas presented so far can explain the atmosphere of **conflict** between science and religion. For example, science proposes the theory of Big Bang about origin of universe, while religious books teach the creation story under various versions out of which the book of genesis of Bible is the most famous. Similarly, miracles are part of religious beliefs; science vehemently refutes the possibility of miracles. The discipline of science deals exclusively with the material objects having physical cause-effect relations; it is a physical reduction of the natural world. On the other hand, religion conceives the world in two realms, namely the supernatural beings and the natural phenomena, assuming that the former controls the affairs of the latter.

This divergence between science and religion puts us in a dilemma about the truth of knowledge in these two fields. We often ask: Why do science and religion give different ideas about the same topic of world and life? What is true, science or religion? How can we reconcile science and religion so as to explain their co-existence? Due to space constraints, this section does not delve into the social, historical and epistemological dimensions of the problem. [# 2]

The philosophical arguments for solving the issue have been already developed in this book. Specifically, the *system model of knowledge*, as reproduced below, can be at hand for synthesizing the disparate knowledge of science and religion.

Methodology: In any discipline there are five kinds of propositions designated as Theory (Ty), Hypothesis (H), Deduction (D), Testing (T) and inference (I), which are ordered like the organs of an animal. Technically speaking, the laws about cause-effect relations and other sensible properties of things constitute our knowledge. But such inferences are reached through the successive stages of Ty, H, D, T and I. Therefore, as per the epistemology or theory of knowledge of System Philosophy, **the most important assertion** is that *all kinds of knowledge under various disciplines have the same methodological scheme denoted by TyHDTI*. This fact synthesizes science and religion, which have different sets of theories and subsequent stages of methodology, as shown in the following Table.

Table : Science-Religion Synthesis [*]

	Science		Religion	
	Content view	**Process view**	**Content view**	**Process view**
Rational view	Abstract concepts of classical science TyHD (1)	Abstract concepts of modern science TyHD (2)	Abstract concepts of theist religion TyHD (3)	Abstract concepts of nontheist religion TyHD (4)
Empirical view	Sensory experience and analysis: TI (1)	Sensory experience and analysis: TI (2)	Sensory experience of worship and analysis: TI (3)	Sensory experience of worship and analysis: TI (4)

Seeing science and religion as two parallel systems of knowledge with the same methodological scheme is just the first step of unification. For completing our pursuit for synthesis, it is necessary to solve the problems of source, justification and truth also as following.

Source: As per *system philosophy of mind*, human mind is a system formed by the dual aspects of *brain and other parts of nervous system (BNS)* and *consciousness*. Then human mind has many levels or faculties, which are mainly classified into scientific mind, religious mind, artistic mind and philosophical mind. When we consider a particular faculty,

BNS refers to the specific area of brain with corresponding neuron networks. At present we are focusing on scientific mind and religious mind. Each of these faculties is like a factory for producing two classes of propositions called deductive propositions (TyHD) and inductive propositions (TI). The ontological existence of the different faculties of mind is the key principle for synthesizing science and religion.

Justification: The philosophical enquiry about the existence of the objects that give various kinds of experiences – the objects of knowledge - is called justification. A proposition about a thing or event belonging to the universe is produced by combining sensory experience and rational thinking in a particular proportion. In order that such a proposition is valid, the corresponding thing or event must exist as a part of universe.

We have already established that physical world exists as the system of matter and energy; it has existence by virtue of the system model X-Y coordinate system. Moreover, physical world is the physical reduction of Natural Life System (NLS). Coming to the field of religion, it has been convincingly explained that the existence of Religious Life System (RLS) is the justification for the mystical propositions about the definition of God and other aspects of religious worship. Going backwards, the NLS, RLS and other five life systems exist by the structure of corresponding faculties of human mind.

Truth: The principle of *system model of truth* is used for unifying scientific truth and religious truth. This topic will be elaborated in the next book, *Discovery of Reality*.

From the foregoing, we may summarize: *System Philosophy shows the synthesis of science and religion by treating them as two parallel systems of knowledge, which are two levels in terms methodology, source, justification and truth.* More importantly, the unification of science and religion is achieved because these are two kinds of knowledge formed by two separate faculties – scientific mind and mystic mind respectively -- of human mind. In the layered view of universe, the Ultimate Reality or *paramporul* is an X-Y coordinate system of body and consciousness. It causes the formation of a hierarchy of things – including inanimate and living things – in the universe. The existence of human mind with various levels of faculties is explained in this view.

NOTES of Chapter 6

#1   System Philosophy offers the exclusive method to counter the atheist arguments of Richard Dawkin's book, *The God Delusion*.

2   After 1950, there is a proliferation of writings about science-religion problem under the views of both theism and process theology. The main reference books regarding this field are Griffin (2000), Haught (2000), Haught (2001) and Robert John Russell (Editor) (2004). In the history of science-religion conflict, the hostile attitudes of Roman Catholic Church towards Galileo (1564-1642) and Charles Darwin (1809-1882) are famous episodes. However, the synthesis of science and religion will be achieved innovatively in this section by the total integration of traditional philosophies of idealism, materialism and process thought, without resorting to realism.

Bibliography

Alan Grafen and Mark Ridley (2007). *Richard Dawkins: How a Scientist Changed the Way We Think,* (Oxford University Press, Paperback Edition)

Alex Inkeles (1993), *What is Sociology? – An Introduction to the Discipline and Profession*, (Prentice Hall of India Limited, New Delhi, Tenth Indian Reprint, 1993)

Ammar Al Chalabi, Martin R. Turner and R. Shane Delamont (2007), *The Brain – A Beginner's Guide*, (One World- Oxford, England, First South Asian Edition, 2007)

Anthony Harrison– Barbet (1990), *Mastering Philosophy*, (Macmillan, London, 1990).

Armstrong, Karen (1998), *A History of God,* (Arrow Books, London, 1998)

Augustine Perumalil (2000), *Critical issues in the Philosophy of Science and Religion*, (Indian Institute of Science and Religion, Pune and ISPCK, Delhi, 2006).

Behe, Michael J. (1996), *Darwin's Black Box: The Biochemical Challenge to Evolution*, (New York: Touchstone Books, 1996).

Bird, Alexander (2003), *Philosophy of Science,* (Routledge, London, Indian Reprint, 2003).

Beiser, Arthur (2002), *Concepts of Modern Physics,* (Tata McGraw-Hill, New Delhi, Sixth Edition, Second Reprint, 2002)

Blackburn, Simon (1996), *The Oxford Dictionary of Philosophy,* (Oxford University Press, 1996)

Brennan, James F. (2005), *History and Systems of Psychology,* (Pearson education, Delhi, first Indian reprint, 2005)

Brooke Noel Moore and Kenneth Bruder(2005), *Philosophy - The Power of Ideas,* (Tata McGraw-Hill Publishing Co. Ltd, New Delhi, sixth edition, 2005.)

Capra, Fritjof (1983), *The Turning Point,* (Flamingo, London, 1983).

Capra, Fritjof (1992), *The Tao of Physics*, (Flamingo, London, Third edition, 1992)

Capra, Fritjof (1997), *The Web of Life*, (Flamingo, London, 1997).

Caputo, John D (2013), *Truth – Philosophy in Transit*, (Penguin books, 2013)

Chardin, Teilhard de (1965), *The Phenomenon of Man*, (Harper Torchbook Edition, New York, 1965)

Chatterjee, Margaret (1988), *Philosophical Enquiries*, (Motilal Banarsidas, Delhi, 1988)

Copleston, Frederick S.J. (1994), *A History of Philosophy. Vol. I – IX*, (Image Books, Doubleday, 1994).

Darwin, Charles (1859), *On The Origin of Species*, (Dover Edition, New York, 2006)

Davies, Brian (2000), *Philosophy of Religion*, (Oxford University Press, contains the reprint of Hume's article 'Of Miracles', 2000)

Davies, Paul (1995), *About Time – Einstein's Unfinished Revolution*, (Penguin Books, 1995)

Davies, Paul (2007), *Cosmic Jackpot – Why Our Universe is Just Right for Life*, (Houghton Mifflin Company New York, 2007).

Dawkins, Richard (1976), *The Selfish Gene*, (Oxford University Press, Oxford and New York, 1976)

Dawkins, Richard (2007), *The God Delusion*, (Black Swan, Transworld Publishers. London, 2007)

Dawkins, Richard (2009), *The Greatest Show On Earth: Evidence for Evolution*, (Bantam Press, 2009)

Dennett, Daniel (1991), *Consciousness Explained*, (Boston: Little, Brown, 1991).

Esposito, John, et al (2008), *Religion & Globalization: World Religions in Historical Perspective.* (Oxford University Press, New York, 2008)

Ewing A. C. (1994), *The Fundamental Questions of Philosophy*, (Allied Publishers Limited, New Delhi, 1994)

Feser, Edward (2006), *Philosophy of Mind*, (Oneworld, Oxford, 2006).

Francis Abraham (1993), *Modern Sociological Theory*, (Oxford University Press, New Delhi, Ninth Impression, 1993).

Frost. S. E (1989), *Basic Teachings of the Great Philosophers*, (Anchor Books, Doubleday, New York, 1942/1989)

George Thomas White Patrick (1978), *Introduction to Philosophy,* (Surjeet publications, Delhi, 1978).

Grayling A.C. (Editor) (1995), *Philosophy: A Guide Through The Subject,* (Oxford University Press, London, 1995).

Grayling A.C. (Editor) (1998), *Philosophy 2 : Further Through The Subject,* (Oxford University Press, London, 1998).

Green, Brian (2005), *The Fabric of the Cosmos,* (Wintage Books, New York, 2005)

Gribbin, John (2008), *The Universe: A Biography,* (Penguin Books, London, 2008)

Griffin, David Ray (2000), *Religion and Scientific Naturalism,* (State University of New York, 2000)

Grolier Encyclopedia of Knowledge, Volumes 1 – 20, (Grolier Incorporated, Danbury, Connecticut, 1993).

Guttman, Burton (2007), *Evolution – A Beginner's Guide,* (One World - Oxford, England, First South Asian Edition, 2007)

Guttman, Griffiths, Suzuki and Cullis (2006), *Genetics– A Beginner's Guide,* (One World- Oxford, England, First South Asian Edition, 2006)

Guyer, Paul (2008), *Kant,* (Routledge, London and New York, 2006; first Indian Reprint, 2008)

Haught, John F. (2000), *God After Darwin – A Theology of Evolution,* (Westview Press, USA, 2000)

Haught, John F. (2001), *Responses to 101 Questions on God and Evolution,* (Paulist Press, USA, 2001)

Hawking, Stephen W. (1995), *A Brief History of Time,* (Bantam Books, 1995 edition)

Hawking, Stephen W. (2011), *The Grand Design,* (Bantam Books, 2011 edition)

Heil, John (2003), *Philosophy of Mind,* (Routledge, London and New York; Indian reprint, 2003),

Hicks, John (1979), *Causality in Economics,* (Basil Blackwell Oxford, Great Briton, 1979)

Hick, John H. (1994), *Philosophy of Religion,* (Prentice Hall of India Pvt. Ltd, fourth edition, 1994).

Hospers, John(1997), *An Introduction to Philosophical Analysis,* (Allied Publishers Limited, Mumbai, 1997; Original publication by Prentice–Hall in 1953)

Ivor Leclerc (1958), *Whitehead's Metaphysics,* (George Allen and Unwin Ltd, London, 1958).

James.T.Shipman, Jerry. L. Adams and Jerry .D. Wilson (1987), *An Introduction to Physical Science*, (D.C. Heath and company 1987).

Jantsch, Erich (1989), *The Self-organizing Universe*, (Pergamon Press, 1989).

Job Kozhamthadam (editor) (2002), *ContemporaryScience and Religion in Dialogue-Challenges and Opportunities,* (ASSR Publications, Jnana-Deepa Vidyapeeth, Pune, 2002),

Job Kozhamthadam (editor) (2003), *Science, Technology and Values,* (ASSR Publications, Jnana-Deepa Vidyapeeth, Pune, 2003),

Job Kozhamthadam (editor) (2004), *Religious Phenomena in a World of Science,* (ASSR Publications, Jnana-Deepa Vidyapeeth, Pune, 2004),

Job Kozhamthadam (editor) (2005), *ModernScience, Religion and The Quest for Unity,* (ASSR Publications, Jnana-Deepa Vidyapeeth, Pune, 2005),

Kant, Immanuel (2003), *Critique of Pure Reason,* (translated by J. M.D. Meiklejohn, Dover Publications, New York, 2003)

Kuhn, Thomas (1970), *The Structure of Scientific Revolutions,* (University of Chicago Press, 1970)

Kukla, Andre (1998), *Studies in Scientific Realism,* (Oxford University Press, 1998)

Ladyman, James (2002), Understanding Philosophy of Science, (Routledge, London, 2002)

Lavin.T. Z. (1989), From Socrates to Sartre, (Bantam Books, New York, 1989)

Leahey, Thomas Hardy (2005), *A History of Psychology – Main currents in Psychological Thought*, (Pearson education, Delhi, first Indian reprint, 2005)

Levin, William C (1984), *Sociological Ideas,* (Wadsworth Publishing Company, California, 1984)

Lewens, Tim (2007), *Darwin,* (Routledge, London and New York, 2007)

Lipsey, Richard (1983), *An Introduction to Positive Economics,* (ELBS Edition, 1983).

Lucy Mair (1992). *An Introduction to Social Anthropology,* (Oxford University Press, New Delhi, Seventh Impression,1992)

Luke George (2004), *Saptaloka Darshanam- Samgraham,* (PGL Books, Changanachery, Kerala, 2004, in Malayalam language).

Luke, George (2015), *Jeevanum Parinamavum*, (PGL Books, Changanachery, Kerala, 2015, in Malayalam language).

Macquarrie, John (1985), *In Search of Deity – An Essay in Dialectical Theism*, (Cross road Publishing, New York, 1985)

Martin Curd and J. A. Cover (1998), *Philosophy of Science: The Central Issues*, (W. W. Norton & Company, New York / London,1998)

Martin Hollis (2000), *The Philosophy of Social Science*, (Cambridge University Press, First Indian Paperback Edition, 2000)

Masih .Y. (1995), *Introduction to Religious Philosophy*, (Motilal Banarsidas, Delhi, reprint 1995)

Maslin K.T. (2001), *An Introduction to the philosophy of Mind*, (polity Press, UK & USA, 2001)

Max Charlesworth (2006), *Philosophy and Religion- From Plato to Postmodernism*, (Oneworld, Oxford, First South Asian Edition, 2006)

Mayr, Ernst (1999), *This Is Biology*, (Universities Press India Limited, Hyderabad)

McGinn, Colin (1998), *The Character of Mind*, (Oxford University Press, 1998

Michael Haralambos and Robin Heald (1990), *Sociology - Themes and Perspectives*, (Oxford University Press, New Delhi, Tenth Impression,1990)

Michio Kaku and Jennifer Thompson (2007), *Beyond Einstein: The Cosmic Quest for the Theory of the Universe*, (Oxford University Press, New Delhi, 2007)

Miller, James B. (Editor) (2001), *An Evolving Dialogue-Theological and scientific Perspectives on Evolution*, (Trinity Press International, Harrisburg, Pennsylvania, 2001)

Newton, Roger (2010), *The Truth of Science – Physical Theories and Reality*, (Viva books, New Delhi, reprint 2010)

O'Leary, Denyse (2004), *By Design or by chance?* (Augsburg Books, Minneapolis, 2004).

Panda N. C. (1999), *Maya in Physics*, (Motilal Banarsidas, Delhi, reprint 1999)

Parthasarathy A. (2000), *Vedanta Treatise*, (Vedanta Life Institute, Mumbai, 2000)

Robert John Russell (Editor) (2004), *Fifty years in Science and Religion – Ian G. Barbour and his Legacy*, (Ashgate Publishing Ltd, England and USA, 2004),

Robert John Russell, Nancey Murphy and C. J. Isham (Editors) (1999), *Quantum Cosmology and the Laws of Nature : Scientific Perspectives of Divine Action*, (Vatican Observatory Publications, Vatican City State and The Center for

Theology and the Natural Sciences, Berkeley, California; second edition, 1999)

Romijn, Herms (2002), *Are Virtual Photons the Elementary Carriers of Consciousness?* (Journal of Consciousness Studies, 9, No.1, 2002, pp 61-81).

Rosenberg, Alex (2000), *Philosophy of Science, a Contemporary Introduction*, (Routledge, London and New York, 2000),

Russell, Bertrand (1992), *The Problems of Philosophy*, (Oxford University Press, 1992).

Sarojini Henry (2009), *Science Meets Faith*, (St. Pauls Mumbai 2009)

Schilpp, Paul Arthur (Editor) (1941), *The Philosophy of Alfred North Whitehead*, (North Western University, 1941)

Schmidt, Paul F. (1967), *Perception and Cosmology in Whitehead's Philosophy*, (Rutgers University Press, New Jersey, 1967)

Sen, Amartya (1990), *On Ethics and Economics*, (Oxford University Press, Delhi, 1990)

Shaffer, Jerome A. (1994), *Philosophy of Mind*, (Prentice Hall of India, New Delhi, 1994)

Sheldrake, Rupert (2013), *The Science Delusion*, (Coronet, Hodder & Stoughton Ltd, London, 2013)

Sherburne, Donald W. (editor) (1965), *A Key to Whitehead's Process and Reality*, (Indiana University Press, London, 1965)

Smolin, Lee (1998), *The Life of Cosmos*, (Phoenix paperback edition, Great Britain,1998)

Smolin, Lee (2008), *The Trouble with Physics*, (Penguin Books, London, 2008)

Solso, Robert L. (2005), *Cognitive Psychology*, (Pearson education, Delhi, second Indian reprint, 2005)

Sweet, William (2003), *Religion, Science and Non-science*, (Dharmaram Publications, Bangalore, 2003)

Tarnas, Richard (1991), *The Passion of the Western Mind*, (Pimlieo, London, 1991),

Taylor, Richard (1994), *Metaphysics*, (Prentice- Hall of India Pvt.Ltd, New Delhi, 1994)

Teilhard Chardin (1999), *Phenomenon of Man*, (Sussex Academic Press, 1999).

Thilly, Frank (2000), *A History of Philosophy*, (SWP Publishers, New Delhi, 2000).

Thomas. A. P, (General Editor) (2012), *Cell and Molecular Biology – The Fundamentals,* (Green Leaf Publications, Kottayam, Kerala, 2012)

Thomson, Mel (1997), *Philosophy of Religion,* (Teach Yourself Books, UK, 1997)

Urmson J.O. and Johathan Ree (1989), *The Concise Encyclopedia of Western Philosophy and Philosophers,* (Unwin Hyman, London, 1989).

Vijayakumaran Nair and Jayaprakash (2007), *Cell Biology Genetics Molecular Biology,* (Academica, Thiruvananthapuram, Fourth Edition)

Washburn, Phil (1997), *Philosophical Dilemmas: Building a Worldview,* (Oxford University Press, New York, 1997)

Whitehead, Alfred North (1978), *Process and Reality,* (Original Edition 1929; Corrected Edition by David Ray Griffin and Donald Sherburne, New York, The Free Press,1978)

Index of Names

Alan Grafen, 43

Ammar Al Chalabi, 43, 105

Anthony Harrison– Barbet, 122

Aristotle (BC 384-322), 8, 33, 37, 41-2, 92, 96-7, 101, 105

Armstrong, Karen, 122

Bateson, Gregory, 106

Behe, Michael J., 43, 56, 57

Beiser, Arthur, 12

Bird, Alexander, 12, 43

Brennan, James, 43, 104, 106

Brentano, Franz (1838-1917), 81

Buffon, George, 48

Capra, Fritjof, 12, 33, 43, 44

Caputo, John D, 122

Chardin, Teilhard de, 19, 43, 47, 80, 106

Copleston, Frederick S.J., 122

Cover J. A., 12

Crick, Francis (1916-2004), 24

Darwin, Charles (1809-1882), 43, 45, 48-58, 64-7, 81, 125-6, 130

Darwin, Erasmus (1731-1802), 48

Davies, Brian, 122

Davies, Paul, 12

Dawkins, Richard (1941-), 43, 52, 60, 64-8, 122, 126

Dembskey, William, 56, 57

Dennett, Daniel, 43

Descartes (1596-1650),70, 77-80, 86, 96

Durkheim (1855-1917), 125

Esposito, John, 122

Feser, Edward, 104-106

Freud, Sigmund (1856-1939), 81, 126

Frost. S. E, 122

Galileo (1564-1642), 130

George Thomas White Patrick, 104, 122

Grayling A.C., 12, 104-106, 122

Green, Brian, 12

Gribbin, John, 12

Griffin, David Ray, 43, 105, 122, 130

Guttman, Burton, 43, 104, 105

Haught, John F., 43, 130

Hawking, Stephen W., 12

Heil, John, 43, 104-106
Hick, John H., 122, 123
Hume, David (1711-1776), 9, 34, 56, 125
James et al, 43
James, William (1842-1910), 81, 105
Jantsch, Erich, 43, 44
Jayaprakash, 43, 44
Jay Gould, Stephen (1941-2002), 53
Jennifer Thompson, 12
Job Kozhamthadam, 43
Jonathan Ree, 12
Jung, Carl (1875-1961), 81
Kant, Immanuel (1724-1804), 33, 43, 79, 89, 90, 110, 111, 122
Khorana, 24
Lamarck (1744-1829), 19, 48, 49, 58
Laszlo, Ervin, 106
Lavin. T. Z., 12, 122, 123
Leahey, Thomas Hardy, 43, 104-6
Leibniz (1646-1716), 79
Lewens, Tim, 43
Luke, George (1953 -), *see* Prologue of this book
Macquarrie, John, 105, 122, 123
Martin Curd, 12
Marx, Karl, (1818-1883), 125
Masih. Y., 122
Maslin K.T., 104-106

Maturana, Humberto, 106
Max Charlesworth, 122
Mayr, Ernst, 43, 44
Mendel, Gregor (1822-84), 50
Michio Kaku, 12
Miller, James B., 43, 86
Newton, Isaac (1642-1727), 20
Newton, Roger, 12
Nietzsche (1844-1900), 125
Nirenberg, 24
O'Leary, Denyse, 43
Paley, William (1743-1805), 56
Penfield, Wilder, 106
Plato (BC 428-348), 78, 114-5, 123
Quine W. V. O, 123
Robert John Russell, 12, 43
Romijn, Herms, 106
Rosenberg, 12, 123
Searle, John, 106
Shaffer, Jerome A., 43
Sheldrake, Rupert, 106
Smolin, Lee, 12
Spencer, Herbert (1820-1903), 50
Spinoza, Benedict (1632–1677), 79
Solso, Robert L., 104-106
Tarnas, Richard, 12, 43, 122
Thilly, Frank, 122
Thomas. A. P, 43
Thomson, Mel, 122

Urmson J.O., 12

Varela, Francisco, 106

Vijayakumaran Nair, 43, 44

Wald, George, 106

Watson, James (1928 -), 24

Watson, John (1878-1958), 82

Whitehead, Alfred North (1861-1947), 19, 47

Wiener, Nobert, 106

Wilber, Ken, 106

Index of Subjects

algorithm, 3, 21, 28, 31-5, 41, 44, 53-5, 66, 83-85

analytic geometry, 37

anthropic, 56, 57

artistic mind, 128

astronomy, 3, 66

A, T, G and C, 24, 25

atheism, 56, 59, 60, 67-8, 111-2, 124-6, 130

behaviorism, 81, 82, 83

Bible, 47, 127

bioinformatics, 31

biotechnology, 30, 31

BNS, *see* nervous system

body, definition of, 11, 14-16

body-mind dualism, 58, 69, 70, 78-9, 86-7, 104

Buddhism, 110

cambrian explosion, 54

cell, definition of, 5, 14-24

cell theory, 22, 30

central dogma, 24, 27-8, 30-1, 44

chance, 49, 51, 55, 60

chariot model of mind, 78, 105

Chinese room, 106

Christian mysticism, 110

chromosome, 23-26, 31, 38

classical biology, 14, 18-21, 34, 41, 47, 51

classical science, 3-9, 32

cloning, 30, 31

cogito ergo sum, 96

cognitive mind, 10, 75-6, 93-96, 105, 114, 118, 121

cognitive psychology, 32, 83, 102

complementary opposites, 8, 9, 37, 44, 58, 62, 90-1, 98, 103, 116, 120

computational / representational theory of thought or CRTT, 84

computer model, 13, 28-33, 41, 44, 53, 66

computer model functionalism (CMF), 7, 32, 77, 81-85, 107

consciousness, 9, 10, 14, 32, 35, 38, 66, 69-106, 118-129

content view, 6, 10, 17-18, 32, 36, 41-2, 54, 58. 61-69, 77, 83, 88, 91, 102-3, 109-118

cosmological puzzle, 11
cosmology, 3-11, 32, 57,
Creationism, 54, 58, 64-68
creative evolution, 55, 68
Darwinism, 45, 49-68, 126
deductive propositions (DP), 5, 9, 29, 118, 121, 129
Diagrams *(of this book)*, 40, 63, 93, 95, 100
dialectical theism, 116, 123
divided line, 114, 115, 123
DNA, 21-42, 52-3, 60, 66, 89-90, 97
double helix, 24, 27
dualism, (*see* body-mind dualism)
eastern mysticism, 19
economics, 37, 44, 68, 112
eliminativism, 80
emergence, 5, 14, 33-4, 41-4, 54, 60, 62, 66, 83, 91
empirical theism, 110, 115, 116, 122
empiricism, 7, 18, 20, 77, 111, 116, 121, 125
enzyme, 27
epiphenomenalism, 34, 41, 77, 80, 85-6
epistemology, 5, 6, 10, 14, 28-9, 58, 108, 117, 127
evolution, definition of, 5, 10, 14, 17, 19, 34, 39, 46-55
existence, 2-11, 21, 32-42, 79-107
existence of God, 108-120

evolutionary theology, 19, 43, 47, 126
evil, existence of, 59, 60, 112, 117-120
FAQ, 120
fossil, 47-55, 61-67
functionalism, 7, 81
functional units of cell, 21-4, 28-41
gene / genetic code, 24-55, 60-90
gene therapy, 30
genetic engineering, 30
genetics, 14, 22, 26-7, 30, 33, 44-57, 89
genome, 26, 31
giraffe, 48
God, existence of, (*see* existence of God)
High Conscious Mind (HCM), 102-104
Hindu mysticism, 110
homunculus, 78-79, 86, 105
Hormone, 24, 27
Human Genome Project, 26, 31
Idealism, 47, 67, 77, 86, 114, 120, 125, 130
identity theory, 81, 82
impressions, 71, 84, 90, 91
inanimate and physical, 96-98
inductive propositions (IP), 5, 9, 29, 129
Inference to the Best Explanation (IBE), 57, 58
information, 24-44, 60-66, 83-90

inheritance, 26, 49-55, 65

input-output model, 21, 83, 103

instrumentalism, 32

intellectual mind, 75-6, 105, 114-116, 119, 123

Intelligent Designer Argument, 39, 45, 55, 58

intentionality, 75, 84, 106

interactionism, 78, 80

Irreducible Complexity Principle. 56-7

isoquant, 60, 61, 65, 68

justification, 6-11, 29-41, 55, 67, 79, 96, 103, 108-121, 128, 129

Kabbalah, 110

knowledge, 2-9, 20, 29-33, 41, 73-79, 90, 108-119, 127-129

Lamarckism, 48

language of thought, 84, 106

layered universe, 30

life, philosophy of, (*see* System Model of life)

life is a puzzle, 22

logical positivism, 7, 30-32, 53-5, 82, 126

Low Conscious Mind (LCM), 102-104

machine-algorithm, (*see* algorithm)

macromolecule, 23-4, 30-43, 53-55, 60-66, 89-90

materialism, 7, 10, 20, 27, 33, 39, 47, 51-67, 80-86, 111-4, 120-130

matter-consciousness system, 10, 66, 90-101

mechanistic worldview, 2, 3, 20

mental states, (*see* System Model of mind)

metabolism, 16, 20-26, 30, 42, 72

methodology, 6, 7, 29, 30, 53, 111-6, 127-129

mind, (*see* System Model of mind)

mind-body dualism, (*see* body-mind dualism)

missing links, 54, 63-68

mitochondria, 23, 27, 31

model, (*see* System Model)

molecular biology, 14, 22-33, 44, 51, 57

Modern Phenomenalism, 7

Mophism, 7

mutation, 50-66

mysticism, 19, 47-8, 60, 68, 93-96, 114, 119, 120

mystic mind, 75-80, 105-123, 129

Mystic Worship Theology, 110, 116

naïve realism., 8, 32

naturalism, 18, 20, 34, 39, 41, 47, 51-60, 77

natural selection, 49-55, 59-66, 126

neo-Darwinism, 52, 64

nervous system, 9, 15, 16, 57, 69-92, 103-5, 128-9

neural networks, 71-74, 83-91, 95-104

neuron, 16, 71-2, 79-89, 122, 129

neuroscience, 71-83, 106, 122

neurotransmitter, 71

OMEGA, 43

ontology, 28, 29, 61

organic worldview, 2, 18, 47, 77, 114

origin of species, 49, 55, 67, 81

origin of universe, 1, 4, 12, 29, 43, 68, 124, 127

panentheism, 110

panpsychism, 77, 79, 105, 140

pantheism, 105, 110, 125, 126

paradigm, 2, 3, 7, 20, 31, 34, 49, 81

parallelism, 77, 79, 105

paramporul, 11, 118, 129

parapsychology, 19, 43

particle-wave duality, 3, 7

phenomenology, 77, 79

philosophic mind, 76

philosophy of religion, 108-123

physicalisation, 88

physical process view, 2-7, 18, 21, 28, 31- 67, 77-85, 97, 106, 126

physical science, 1-19

prana, 18, 19

problem of evil, 59, 119, 120

process theology, 47, 110, 116, 120-130

process view, 18-9, 21, 32

production function, 37, 44, 68, 90, 98

protein, 21-42, 53, 60, 73, 89, 97

Protestant Church, 110

pseudoscience, 19, 43

psychoanalysis, 81, 106

psychology, 32, 70, 71, 78-106

punctuated equilibrium, 53, 54

quantum consciousness, 80, 106

quantum cosmology, 4-9, 32

quantum field theory, 3, 4

quantum mechanics, 3-9, 30-32

quantum mind, 80

quantum physics, 3, 29, 33

rationalism, 7, 18, 77, 78, 111, 116, 121

rational theism, 110, 114-9, 122

realism, 8, 10, 18, 36, 48, 67-87, 97-102, 110, 119-121, 130

reality, 11, 66, 70, 79, 111-129

religion, 108-130

religious life system (RLS), 113-129

religious reality, 117

Renaissance, 19

representationalism, 84

Ribo Nucleic Acid (RNA), 23

Roman Catholic, 110, 130

romanticism, 18, 19, 47

rules of thought, 8, 33, 37, 42, 92-97

science-religion synthesis, 124-130

scientific mind, 6-10, 30, 76, 98, 124-9

scientific realism, 8, 10, 32, 33, 55, 58, 66

scientism, 55, 67, 111

self-consciousness, 72-76, 96, 102-4, 121

selfishness of genes, 52, 55

self-interest, 113, 117

self-organization, 33-5, 41, 44

Seven Life Systems, 112-113

skepticism, 8, 32, 34, 41, 55, 58

social system, 108-113

society-interest, 113, 117

soul, 7, 18, 35, 51, 70, 77-9, 86-7, 96, 108, 114-122

source, 6, 7, 29, 30, 38, 98, 111-9, 121, 126-9

specialization of cells, 15, 24, 35

species, 16-26, 46-70, 81

spiritual process view, 2, 7, 18-9, 35, 43- 49, 58, 68-85, 112-120

stimulus-response, 82

struggle for life, 49-55

sufism, 110

system cosmology, 11

System Model,

    - definition, 36-37
    - of biological world, 8, 42-44, 60-68, 70
    - of evolution, 58-64
    - of life and organism, 40-2, 66, 90, 96, 126
    - of mind, 85-103
    - of physical world, 9-11
    - of reality, 11, 118, 129
    - of religion, 112-123

Tables *(of this book)*, 18, 38, 76, 77, 110, 115, 128

taoism, 110

teleology, 59

theism, 47, 65, 67, 109, 111-130

transcendent, 109, 114

TyHDTI scheme, 5-10, 29, 30, 115-121

Ultimate Reality, *(see* reality)

unconscious mind, 72-84, 91-95, 102-6

underdetermination, 32

verification principle (verifiability criterion of meaning), 7, 30, 53, 82

vitalism, 18, 35, 39, 47, 56, 58, 60

what is life?, 14-17, 22, 41-43

worldview, 2-7, 17-22, 28, 39-47, 60, 77, 81, 85, 94, 105-122

yoga, 18, 19

www.ingramcontent.com/pod-product-compliance
Lightning Source LLC
Chambersburg PA
CBHW031053180526
45163CB00002BA/808